The SGTE Casebook
Thermodynamics at work

The SGTE Casebook
Thermodynamics at Work

Edited by Dr K. Hack
GTT-Technologies, Herzogenrath, Germany

MATERIALS MODELLING SERIES

THE INSTITUTE OF MATERIALS
1996

Book 621
First published in 1996 by
The Institute of Materials
1 Carlton House Terrace
London SW1Y 5DB

ISBN 0 901716 74 X

MATERIALS MODELLING SERIES
Series Editor: Dr H. K. D. H. Bhadeshia
The University of Cambridge
Department of Materials Science
and Metallurgy

Typeset by The Institute of Materials
Printed and bound in the UK by
Bourne Press, Bournemouth

Dedication

This book is dedicated to Professor E. Bonnier in recognition of his enthusiasm, vision and patience in the creation, development and implementation of a European-based structure for the Scientific Group Thermodata Europe, SGTE. His wise leadership was largely responsible for the present availability to the materials industry of a valuable thermodynamic tool for the development and optimisation of materials and processes.

Editor's Acknowledgements

My deep felt gratitude goes to Mats Hillert who supported the book from the first idea through many stimulating discussions, especially on the Gibbs energy modelling, to its finalisation.

I would like to express my thanks to all contributing authors without whom this book never could have been realised.

I would like to thank Robert C. Fullerton-Batten, Hans Kürten and especially my wife for their support with the manuscript.

Klaus Hack

Foreword

The major purpose of this book is to illustrate how thermodynamic calculations can be used as a basic tool in the development and optimisation of materials and processes of many different types.

The examples selected are, to a large extent, real case studies dealt with by members of the Scientific Group Thermodata Europe, SGTE, in the course of their work.

SGTE is a consortium of European laboratories working together to develop high quality thermodynamic databases for a wide variety of inorganic and metallurgical systems. The SGTE data can be obtained via members and their agents for use with commercially available software developed by some of the members, to enable users to undertake calculations of complex chemical equilibria efficiently and reliably. The case studies presented in the book have been treated using SGTE data in combination with such software.

Members of SGTE have played a principal role in promoting the concept of 'computational thermochemistry' as a time and cost-saving basis for guiding materials processing of many different types. In addition, such calculations provide crucial process-related information regarding the nature, amounts and distribution of environmentally hazardous substances produced during the different processing stages.

While further developments in data evaluation techniques, in the modelling of different types of stable and metastable phases, in the coupling of thermodynamics and kinetics and in the scope of application software are still needed, the case studies presented in this volume demonstrate convincingly that thermochemical calculations have very great potential for providing a sound and inexpensive basis for materials and process development in many areas of technology.

The book is dedicated to Professor E. Bonnier, the first Chairman of SGTE, whose vision, continuous effort, patience and guidance through the formative years of SGTE as a European group, first as a project supported by the French CNRS and afterwards by DG XIII of the European Community, were the major inspiration for the establishment of the present, wide-reaching joint activities of SGTE members. Apart from its continuous work of thermodynamic data evaluation, SGTE now carries out its own joint research contracts and via its members and agents offers its data world-wide both on-line and in the form of packages for use with main-frames or PC computers. As an organisation, SGTE cooperates in a broader international effort to unify thermodynamic data and assessment methods.

I. Ansara
P. Spencer (Chairman)

SGTE Member Organisations

France:

Institut National Polytechnique (LTPCM), Grenoble

Association THERMODATA, Grenoble

IRSID, Maizières-lès-Metz

Germany:

Rheinisch-Westfälische Technische Hochschule (LTH), Aachen

MPI für Metallforschung (PML), Stuttgart

Sweden:

Royal Institute of Technology (Materials Science and Engineering), Stockholm

United Kingdom:

National Physical Laboratory (MTDS), Teddington

AEA Technology, Harwell

Table of contents

Introduction 13

Part 1: Theoretical Background 17

1 Basic Thermochemical Relationships 17
KLAUS HACK

1.1 Thermochemistry of stoichiometric reactions 18
1.2 Thermochemistry of complex systems 21

2 Models and Data 25
KLAUS HACK
2.1 Gibbs energy data for pure stoichiometric substances 27
2.2 Gibbs energies for solution phases 33

3 Graphical Representations of Equilibria 43
KLAUS HACK

4 Summarising Mathematical Relationships between Gibbs Energy and other Thermodynamic Information 51
KLAUS HACK

Part II: Applications in Materials Science and Processes 53

5 Hot Salt Corrosion of Superalloys 56
TOM I. BARRY AND ALAN T. DINSDALE

6 Computer Assisted Development of High Speed Steels 70
PER GUSTAFSON

7 Using Calculated Phase Diagrams in the Selection of the Composition of Cemented WC Tools with a Co–Fe–Ni Binder Phase 77
ARMANDO FERNÁNDEZ GUILLERMET

8 Prediction of Loss of Corrosion Resistance in Austenitic Stainless Steels 85
MATS HILLERT AND CAIAN QIU

9 Calculation of Solidification Paths for Multicomponent Systems 94
Bo Sundman and Ibrahim Ansara

10 Prediction of a Quasiternary Section of a Quaternary Phase Diagram 99
Mats Hillert and Stefan Jonsson

11 Estimative Treatment of Hot Isostatic Pressing of Al-Ni Alloys 103
Klaus Hack

12 The Thermodynamic Simulation in the Service of the CVD Process. Application to the Deposition of WSi_2 Thin Films 108
Constantin Vahlas, Claude Bernard and Roland Madar

13 Calculation of the Concentration of Iron and Copper Ions in Aqueous Sulphuric Acid Solutions as a Function of the Electrode Potential 118
Jürgen Korb and Klaus Hack

14 Thermochemical Conditions for the Production of Low Carbon Stainless Steels 129
Klaus Hack

15 Application of Phase Equilibrium Calculations to the Analysis of Severe Accidents in Nuclear Reactors 135
Richard G. J. Ball, Paul K. Mason and Mike A. Mignanelli

16 Pyrometallurgy of Copper-Nickel-Iron Sulphide Ores: The Calculation of Distribution of Components between Matte, Slag, Alloy and Gas Phases 151
Tom I. Barry, Alan T. Dinsdale, Susan M. Hodson, Jeff R. Taylor

17 High-Temperature Corrosion of SiC in Hydrogen-Oxygen Environments 163
Klaus G. Nickel, Hans L. Lukas and Günter Petzow

18 The Carbon Potential during Heat Treatment of Steel 176
Torsten Holm and John Ågren

19 Preventing Clogging in a Continuous Casting Process 183
Bo Sundman

20 Evaluation of EMF from a Potential Phase Diagram for a Quaternary System 187
MATS HILLERT

Part III: Towards Process Simulation 191

21 Steady-State Calculations for Dynamic Processes 193
KLAUS HACK

22 Diffusion in Multicomponent Phases 196
JOHN ÅGREN

23 Production of Metallurgical Grade Silicon in an Electric Arc Furnace 200
GUNNAR ERIKSSON AND KLAUS HACK

24 Multicomponent Diffusion in Compound Steel 209
JOHN ÅGREN

Name Index 215

Subject Index 218

Introduction

> The real *raison d'être* for the continuation of extensive experimental research in metallurgical thermochemistry is the potential application of its principles and data to practical, in particular industrial, problems. For this purpose the gathering of raw experimental data is obviously not enough. Missing numerical information must be supplemented by estimates ... Raw data must be sifted and critically evaluated to provide for every chemical system a consistent set of thermochemical properties. ...
> In practice, it is true, the knowledge of reaction rates is as important as that of equilibrium, if not more so, but the kinetic problems can only be tackled when the thermodynamic ones have been settled. It is also true that, in practice, metallurgical reactions are quite involved ... but with some effort it will be found that even complicated chemical processes may be broken up into simpler reactions which are accessible to normal thermodynamic evaluation.

The above points are made in the 5th edition of *Metallurgical Thermochemistry* by Kubaschewski and Alcock in 1979 [79Kub]. Elsewhere in the same book the term data-bank is used, albeit in quotation marks. Most of the statements are still relevant: computer supported calculations provide an enormous potential for the application of thermodynamic principles to the solution of practical problems. There is still the need for good estimates arising from the lack of data in certain fields of interest; and critical evaluation of raw experimental results to obtain consistent thermodynamic data sets for complete chemical systems is still of paramount importance. Nevertheless, the development of software for treating thermochemical problems has made some considerable advances in the past two decades and the questions that remain open can be tackled in a much more comprehensive way.

The enormous effort involved in data collection and evaluation as carried out for example by Kubaschewski for pure substance data and by Kaufman [78Kau] in the field of alloy phases is now a somewhat less arduous task due to the availability to thermochemists of the computer. This has made it possible to treat thermochemistry in a completely new way. The computer, because of its data storage and management **and** its 'number-crunching' capabilities, has enabled us to look at the thermochemistry of a system as a whole, i.e. in many cases the user needs nothing more than a list of elements in his system and the values of the global variables temperature, pressure and element concentrations to carry out a theoretical study. Calculations can then be made of the phases stable at equilibrium, their amounts and compositions, and even information about the degree of instability of the phases not present at equilibrium can be provided. The flow sheet shown in Figure I may be used to illustrate the work procedure entailed in their activities.

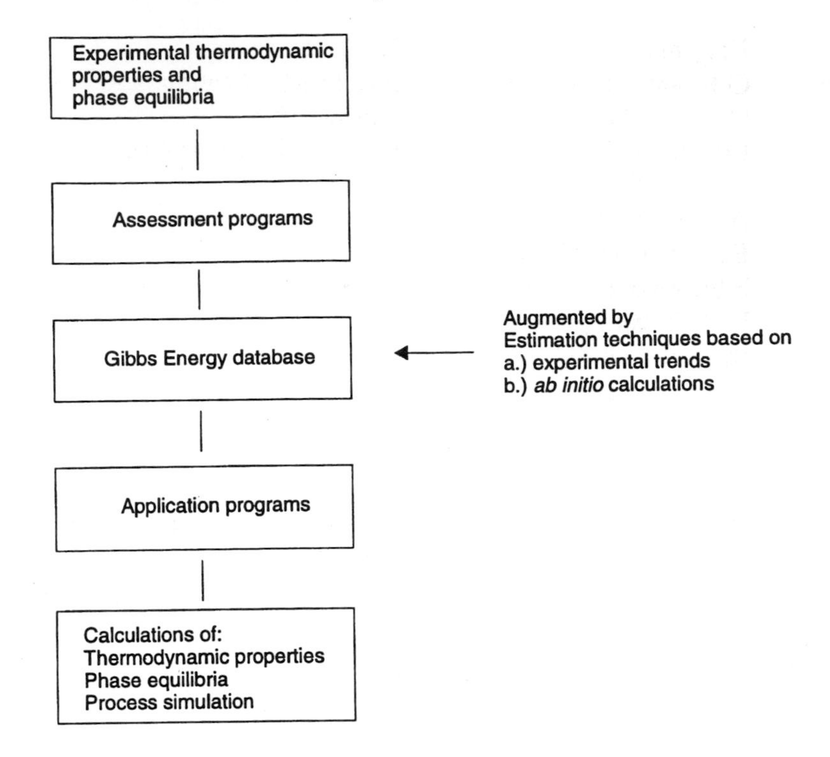

Fig. I.1 Flow sheet of the work procedure, from data assessment to an application calculation.

The purpose of the present volume is to present some examples of such calculations and thus to demonstrate the enormous potential of this new technique. The computerised databases are still quite limited but a considerable effort is underway to expand them.

SGTE is making a major effort to provide comprehensive high-quality self-consistent, computerised thermodynamic databases both for pure substances and mixtures of all types and is playing a leading role in establishing methods for data evaluation and modelling of solution phases. Software for the storage and retrieval of assessed data has been developed and there are a number of application programs to treat different aspects of chemical equilibrium [70Kau, 80Bar, 83Tur, 84Sch, 85Sun, 85Tho, 85Tur, 87Bar, 88Che, 88Din, 88Roi, 88Sun, 88Tho].

References

70Kau L.Kaufman and H.Bernstein: *Computer Calculation of Phase Diagrams*, Acad. Press, New York, 1970.

78Kau L.Kaufman and H.Nesor: *Calphad: Comput. Coupling Phase*

Diagrams Thermochem. **2**, 1978, 55-80.

79Kub O.Kubaschewski and C.B.Alcock: *Metalurgical Thermochemistry* 5th edn, Pergamon Press, Oxford, 1979.

80Bar I.Barin, B.Frassek, R.Gallagher and P.J.Spencer: *Erzmetall,* **33**, 1980, 226.

83Tur A.G.Turnbull: *Calphad* **7**, 1983, 137.

84Sch E.Schnedler: *Calphad* **8**, 1984, 265-279.

85Sun B.Sundman, B.Jansson and J.-O.Andersson: *Calphad* **9**, 1985, 153.

85Tho W.T.Thompson, A.D.Pelton and C.W.Bale: F*A*C*T Facility for the Analysis of Chemical Thermodynamics, Guide to Operations, McGill University Computing Centre, Montreal, 1985.

85Tur A.G.Turnbull and M.W.Wadsley: The CSIRO-SGTE THERMODATA System, CSIRO, Inst. of Energy and Earth Resources, Port Melbourne, Australia, 1985.

87Bar T.I.Barry, A.T.Dinsdale, R.H.Davies, J.Gisby, N.J.Pugh, S.M.Hodson, and M.Lacy: *MTDATA Handbook: Documentation for the NPL Metallurgical and Thermochemical Databank,* National Physical Laboratory, Teddington, U.K., 1987.

88Che B.Cheynet: Int. Symp. on Computer Software in Chemical & Extractive Metallurgy, *Proc. Metall. Soc. of CIM,* Montreal '88 **11**, 1988, 87.

88Din A.T.Dinsdale, S.M.Hodson, T.I.Barry and J.R.Taylor: Int. Symp. on Computer Software in Chemical & Extractive Metallurgy, *Proc. Metall. Soc. of CIM,* Montreal '88 **11**, 1988, 59.

88Roi A.Roine: Int. Symp. on Computer Software in Chemical & Extractive Metallurgy, *Proc. Metall. Soc. of CIM,* Montreal '88 **11**, 1988, 15.

88Sun B. Sundman,: Int. Symp. on Computer Software in Chemical & Extractive Metallurgy, *Proc. Metall. Soc. of CIM,* Montreal '88 **11**, 1988, 75.

88Tho W.T.Thompson, G.Eriksson, A.D.Pelton and C.W.Bale: Int. Symp. on Computer Software in Chemical & Extractive Metallurgy, *Proc. Metall. Soc. of CIM,* Montreal '88 **11**, 1988, 87.

Part I: Theoretical Background

1 Basic Thermochemical Relationships

Klaus Hack

Since the publication of Gibbs's last paper [878Gib] in the series 'On the equilibrium of Heterogeneous Substances' in 1878, all terms necessary to describe (chemical) equilibrium are defined. The chemical potential had been introduced, and the relation governing the different types of phase diagram (the Gibbs-Duhem equation) had been derived. Furthermore the different work terms in what we now rightly call Gibbs's fundamental equation had been discussed far beyond the contribution of chemical or electrical work and included already, for example the contribution of surface tension or the gravitational potential. Gibbs also stated clearly that it is only the relative magnitude of each of these terms that permits omission for practical purposes, in principle all possible contributions are always present.

Most problems dealt with in equilibrium thermochemistry are those with constant temperature and pressure and where the other work terms, except for the chemical contribution, are usually omitted. Electrochemistry, of course, can only be treated if the electrical work term is also explicitly included. It is important to keep this in mind since the entire database derived under these conditions is a Gibbs energy, rather than a Helmholtz, Enthalpy or Internal energy database. Problems with constant temperature and volume, for example, have thus to be treated in an indirect way, which is, of course, no problem for the computer.

Using the Maxwell-relations one can easily derive a diagrammatic scheme (Figure 1.1) to relate the Gibbs energy in its natural variables ($G(T,P)$) with the other state functions and their natural variables, i.e. the Helmholtz energy, $F(T,V)$, the enthalpy, $H(S,P)$, and the internal energy, $U(S,V)$.

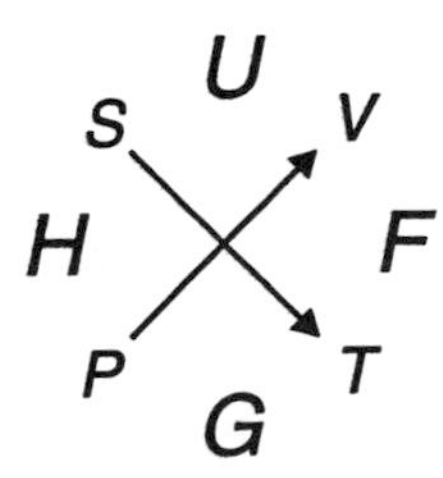

Fig. 1.1 Diagram representing the Maxwell relations.

The arrows in the scheme indicate the signs of the derivatives one has to take of the respective state function with respect to the chosen natural variable, e.g. $(\partial G/\partial P)_T = V$ or $(\partial G/\partial T)_P = -S$. It shall suffice to say here that a complete change from one state function to another can be obtained by application of a mathematical procedure, called the Legendre transformation [71Hit].

Equilibrium is established if the potential function of the system for the conditions chosen has reached an extremum; in the case of the Gibbs energy as function of T, P, and the mole numbers, etc., it is a minimum as expressed by the following equations.

$$G = \min \quad \text{or} \quad dG = 0 \quad \text{and} \quad d^2G \geq 0 \tag{1.1}$$

Gibbs's fundamental equation yields:

$$dG = -SdT + VdP + \Sigma\mu_i dn_i + \Sigma z_j F\varphi_j dn_j \ldots \tag{1.2}$$

with total entropy S, temperature T, total volume V, pressure P, chemical potential μ, molecular number n, charge number z, Faraday constant F, electric potential φ. From (1.1) and (1.2) two different routes for a quantitative approach to equilibrium are possible. These are described in the following two sections.

1.1 Thermochemistry of stoichiometric reactions

The historical route, established experimentally before Gibbs, is the method of stoichiometric reactions. For isothermal and isobaric conditions, disregarding electrical and other work terms in a system, one obtains:

$$dG = \Sigma\, \mu_i dn_i \tag{1.3}$$

The mass balance of a stoichiometric reaction can generally be written as:

$$\Sigma \nu_i B_i = 0 \tag{1.4}$$

with ν being positive for products and neagtive for reactants. Thus the changes dn_i of the absolute mole numbers n_i of the substances B_i are defined by the change dξ of the extent of reaction ξ and the stoichiometric coefficients ν_i of the mass balance equation (1.4).

$$dn_i = \nu_i\, d\xi \tag{1.5}$$

After splitting the chemical potential μ into the reference potential μ° and the activity contribution $RT \ln a$ (R = general gas constant)

$$\mu = \mu^\circ + RT \ln a \tag{1.6}$$

one obtains the well known law of mass action expressed in the equation for equilibrium:

$$\Delta G^\circ = \Sigma \nu_i \mu_i^\circ = -RT \ln \left(\Pi a_i^{\nu_i} \right) = -RT \ln K \tag{1.7}$$

This equation permits the derivation of most informative relations between the activities of the products and reactants:

$$a_i = \text{func}(a_j, T) \quad \text{with } j \neq i \tag{1.8}$$

It should be noted that the temperature dependence of this relationship is contained solely in the Gibbs functions of the pure substances ($\mu_i^\circ = G_i^\circ(T)$) that are involved in the reaction.

However, in practice one is usually interested in a relationship between concentrations rather than activities. The derivation of such a relation based on a stoichiometric reaction approach is perfectly feasible but is subject to two pitfalls, one mathematical, the other a chemical. Firstly, the use of numerical methods cannot be avoided except in the simple case of ideal homogeneous systems e.g. gas equilibria. In general, one has to deal with transcendental equations and even in the simple case an auxiliary equation for the total pressure of the system has to be employed. It is, in other words, not a question of straight linear algebra.

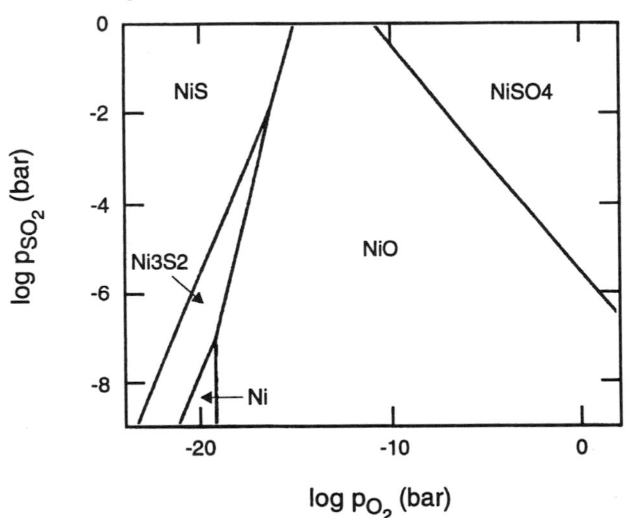

Fig. 1.2 Phase stability diagram for Ni as a function of partial pressure of O_2 and SO_2 at 873 K. Use of this diagram could give misleading impression of dependence of coexistence lines on log p_{O2} at high and low oxygen potentials, cf Fig. 1.3.

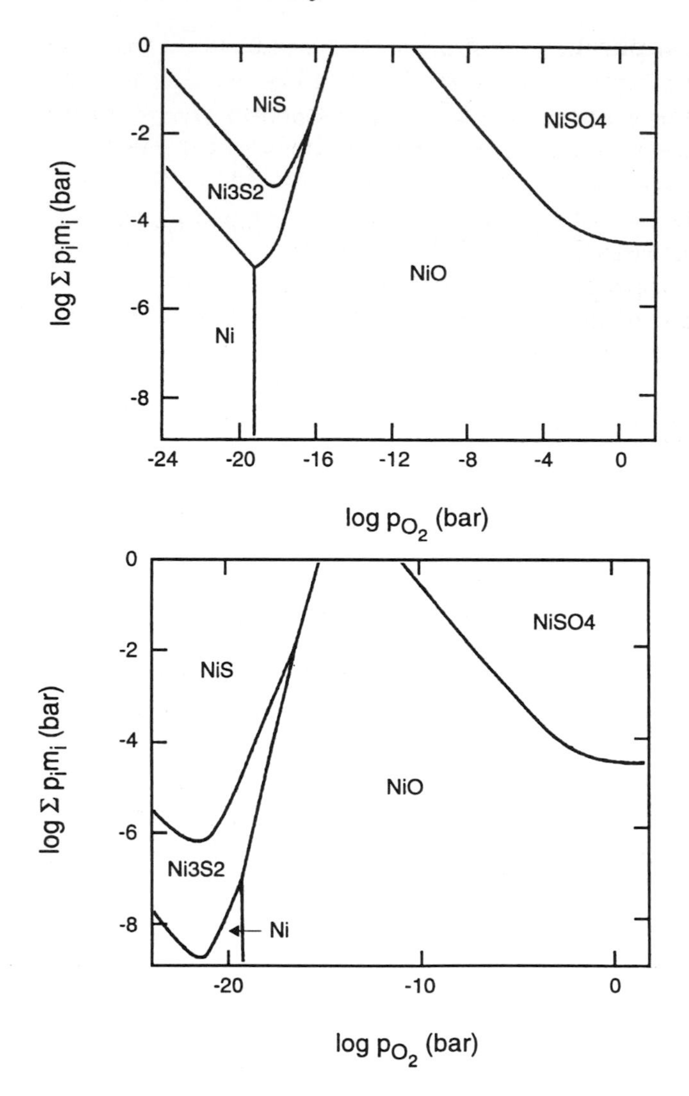

Fig. 1.3 Phase stability diagrams for Ni as a function of log p_{O2} and log $\Sigma p_i m_i$, where p_i is partial pressure of each species containing S and m_i is stoichiometry number of S in species i at 873 K for two different pressures of H_2O. (a) $p_{H2O} = 10^{-5}$ bar; (b) $p_{H2O} = 1$ bar.

Secondly, and much more importantly, **all** the independent reactions in a system must be known **before** starting the calculation. In other words, one must either make assumptions on the complete set of independent reactions in the system, or analyse these experimentally before a reasonable calculation can be carried out. Such assumptions can easily lead to simplifications with very striking differences in the results, for example in a phase diagram. Figures 1.2 and 1.3 show phase stability diagrams for Ni in sulphur- and oxygen-containing atmospheres with additions of H_2O(gas). This is a typical case for the application of stoichiometric reactions in the derivation of an equilibrium diagram. In Figure 1.2 only O_2 and SO_2 are considered to be important gas species, thus leading to the well known straight line phase boundaries, whereas Figures 1.3a and b show the influence of the entire equilibrated gasphase with a fixed potential of H_2O. Comparison of Figures 1.2 and 1.3a shows that for some conditions, low $p(H_2O)$ and oxygen pressures between 10^{-20} and 10^{-4}, the results are in good agreement, but outside the appropriate pressure range for oxygen the behaviour of the phases is quite different. However, if the partial pressure of H_2O is raised to a much higher level (Figure 1.3b), there is very little similarity left between the diagrams 1.2 and 1.3b because of the severe changes in the gas phase. These cannot be taken into account if one bases all reactions on the assumption that SO_2 and O_2 are the only important gas species under all conditions.

1.2 Thermochemistry of complex systems

The above example leads directly to the second method. Equilibria in complex systems, i.e. systems with many components and many phases – some or all of which may be nonideal mixtures – can only be treated safely by minimisation of the total Gibbs energy of the system under some constraints. This requires the compulsory use of numerical methods. As indicated, computer programs for the solution of multivariable, transcendental equation systems have to be developed. Now, the equilibrium condition is written as

$$G = \sum n_i \mu_i = \text{minimum} \tag{1.9}$$

Here the chemical potentials μ_i refer to the entire set of chemical species in the system, no matter whether they are e.g. gas species or aqueous species and thus part of one phase, or condensed stoichiometric substances such as Al_2O_3 or $CaCO_3$ and thus one phase each.

A clearer way to write the same sum is given by putting a greater emphasis on the phases of the system. After all, only the phases can come into equilibrium and some species used for the description of condensed phases might well be artefacts of the model used for the Gibbs energy of the particular phase. Now

$$G = \sum \left(\Sigma n_i^{\varphi}\right) G_m^{\varphi} = \text{minimum} \tag{1.10}$$

is used. Here, G_m^{φ} is the molar integral Gibbs energy of phase φ and n_i^{φ} are the mole numbers of the phase constituents i of this phase. Thus the inner sum refers to the respective phase amounts.

The mass-balance equations

$$\Sigma n_i a_{ij} = b_j \quad (j = 1 \text{ to } m) \tag{1.11}$$

are subsidiary conditions. n_i is the mole number of species i, a_{ij} are the stoichiometric coefficients of species i with respect to the components of the system (j = 1 to m), and b_j is the total amount of component j. Note that the components of the system are normally but not necessarily the constituent elements. Such an equation system can be solved by the introduction of Lagrangian multipliers M_j. One obtains at equilibrium the simple relationship:

$$G = \Sigma b_j M_j \tag{1.12}$$

It turns out that M_j are the chemical potentials of the system components *at equilibrium*. Note, that there are usually far more species i than system components j ($n >> m$) in a system. As an example, the matrix a_{ij} for a gas-metal-slag system with the elementary components Fe–C–Ca–Mg–N–O–Si is given in Table 1.1.

It is also interesting to note, that equation (1.12) is a simple linear equation which defines the tangential hyper-plane of dimension m–1 to the transcendental Gibbs-energy surface of the system of dimension m. The common tangent rule can easily be derived from it.

At equilibrium, the chemical potentials μ_i of the species can all be calculated from :

$$\mu_i = \Sigma a_{ij} M_j \tag{1.13}$$

It is clear from the above that methods for calculation of complex chemical equilibrium require, on the one hand, models for the description of the molar integral Gibbs energy of ideal and non ideal mixture phases and, on the other, robust and reliable numerical algorithms to solve the constrained equation systems indicated.

From equation (1.10), it is obvious that for each phase an expression for G_i^φ, the integral molar Gibbs energy of the phase φ is required. Two cases can occur: either the phase is treated as a pure stoichiometric substance (compound), e.g. alumina with the formula Al_2O_3, or the phase is a solution (mixture) with variable content of its phase constituents, e.g. a body centered cubic (bcc) alloy of iron and chromium, $\{Fe,Cr\}_{bcc}$. In the first case, the Gibbs energy only needs to be known as a function of T and P, $G(T,P)$, whereas the second case requires the Gibbs energy to be known as a function of T, P and the mole numbers n_i of the phase constituents, $G(T,P,n_i)$. It must be noted that for the modelling of the molar Gibbs energy it is preferable to use

Table 1.1 Example of a stoichiometric matrix for the gas–metal–slag system Fe–N–O–C–Ca–Si–Mg.

Phase	Components	System Components						
		Fe	N	O	C	Ca	Si	Mg
Gas	Fe	1	0	0	0	0	0	0
	N_2	0	2	0	0	0	0	0
	O_2	0	0	2	0	0	0	0
	C	0	0	0	1	0	0	0
	CO	0	0	1	1	0	0	0
	CO_2	0	0	2	1	0	0	0
	Ca	0	0	0	1	0	0	0
	CaO	0	0	1	0	1	0	0
	Si	0	0	0	0	0	1	0
	SiO	0	0	1	0	0	1	0
	Mg	0	0	0	0	0	0	1
Slag	SiO_2	0	0	2	0	0	1	0
	Fe_2O_3	2	0	3	0	0	0	0
	CaO	0	0	1	0	1	0	0
	FeO	1	0	1	0	0	0	0
	MgO	0	0	1	0	0	0	1
Liq. Fe	Fe	1	0	0	0	0	0	0
	N	0	1	0	0	0	0	0
	O	0	0	1	0	0	0	0
	C	0	0	0	1	0	0	0
	Ca	0	0	0	0	1	0	0
	Si	0	0	0	0	0	1	0
	Mg	0	0	0	0	0	0	1

concentrations rather than absolute mole numbers. However, as will be demonstrated in Chapter 2 the choice of the concentration variable, e.g. mole fraction, site fraction, equivalent fraction, is already intimately related to the Gibbs energy model used in a particular case. It may, therefore, suffice here to indicate that, in general, for solution phases the modelling requires the Gibbs energy to be described with at least the following three explicit terms:

$$G_m(T, P, n_i) = G^{ref}(T, P, n_i) + G^{id}(T, n_i) + G^{ex}(T, P, n_i) \qquad (1.14)$$

The first term contains the contribution of the pure phase constituents, the second term gives the contribution due to the ideal mixing of the chosen phase constituents, and the third term contains non-ideal (excess) contributions with respect to the chosen ideal mixing. For an overview of the most widely used Gibbs energy models and their mathematical representation see Chapter 2.

References

878Gib J.W.GIBBS: *Trans. Conn. Acad.* **3**, 1878, 176.

71Hit O.HITTMAIR and G.ADAM: *Wärmetheorie,* Vieweg, Braunschweig, 1971.

2 Models and Data

Klaus Hack

The assessment of the thermodynamic properties of individual phases, i.e. their Gibbs energy as a function of temperature, composition and possibly pressure, is the basis for the successful establishment of a thermodynamic databank. Gibbs energy data have to be made available for phases with a wide variety of properties such as:

Pure stoichiometric phases,
e.g. metallic elements, stoichiometric oxides or gas species in their standard state

Pure stoichiometric condensed substances under high pressure,
e.g. real or synthetic geological phases or pure metals

Ferro-, antiferro- or paramagnetic pure substances
e.g. magnetic elements or oxides

Species forming solutions,
e.g. ideal or non-ideal gases or aqueous solutions

Condensed substitutional solutions,
e.g. alloys with metallic components

Interstitial solutions,
e.g. carbon and gases in alloys

Solutions exhibiting chemical defects,
e.g. non-stoichiometric oxides or salts with differently charged ions

Solutions with several sublattices,
e.g. alloys with metallic components occupying different sites of a lattice or solid salts with equally charged ions

Solutions exhibiting ordering transformations,
e.g. magnetically or chemically ordered alloys with metallic components

The following chapter will give an overview of models used in the assessment of the Gibbs energies of such phases. Once the Gibbs energies are known all other thermodynamic properties can be derived such as:

Volume $$V = \left(\frac{\partial G}{\partial P}\right)_{T,n} \tag{2.1}$$

Entropy $$S = -\left(\frac{\partial G}{\partial T}\right)_{P,n} \tag{2.2}$$

Enthalpy $$H = G - T\left(\frac{\partial G}{\partial T}\right)_{P,n} \tag{2.3}$$

Internal energy $$U = G - T\left(\frac{\partial G}{\partial T}\right)_{P,n} - P\left(\frac{\partial G}{\partial P}\right)_{T,n} \tag{2.4}$$

Helmholtz energy $$F = G - P\left(\frac{\partial G}{\partial P}\right)_{T,n} \tag{2.5}$$

Heat capacity at constant pressure $$C_p = -T\left(\frac{\partial^2 G}{\partial T^2}\right)_{P,n} \tag{2.6}$$

Heat capacity at constant volume

$$C_v = -T\left[\left(\frac{\partial^2 G_m}{\partial T^2}\right)_{P,n} + \left[\left(\frac{\partial^2 G_m}{\partial P \partial T}\right)^2 / \left(\frac{\partial^2 G_m}{\partial P^2}\right)\right]_{T,n}\right] \tag{2.7}$$

Chemical potential (Partial Gibbs energy) $$\mu_i = \left(\frac{\partial G}{\partial n_i}\right)_{T,P,nj} \tag{2.8}$$

We know about the thermodynamic properties of phases through information gathered by experience or, more often, by intentional experiments. This information is formed into a physical picture or model, e.g. oscillating atoms on lattice sites with a restricted number of degrees of freedom to describe the maximum value of the heat capacity. To be able to quantify the measured properties, the next step is to put the physical model into a mathematical form which can subsequently be used for making interpolations or even predictions. Before accepting the prediction given by a model, it is important to test the model by comparing a number of such predictions with experimental information already available or obtained by new experiments. Even after successfully passing such a test, it must be realised that the physical model behind the mathematical description may not give a correct picture of the real world, e.g. the description of carbon in iron or other metals as a substitutional solution. The mathematical description may thus result in incorrect predictions outside the range in which it has been tested.

2.1 Gibbs energy data for pure stoichiometric substances

Although the Gibbs energy is the central function, it is still customary and useful to store and apply the data of a pure stoichiometric substance in the form of the enthalpy of formation and entropy at standard conditions (T=298.15 K and P_{tot}=1 bar) as well as a temperature function of the heat capacity. The latter is integrated over temperature to derive the temperature dependence of H and S.

$$H = H^{ref} + \int_{T_{ref}}^{T} C_p \, dT \tag{2.9}$$

and

$$S = S^{ref} + \int_{T_{ref}}^{T} \frac{C_p}{T} \, dT \tag{2.10}$$

The Gibbs energy is then calculated from the Gibbs–Helmholtz relation $G = H - TS$.

Solid state physics shows that the temperature dependence of the heat capacity C is best explained by a quantum mechanical picture of lattice vibrations [07Ein, 12Deb]. Thus one obtains the Debye function

$$C = D\left(\frac{\Theta}{T}\right) \tag{2.11}$$

where Θ is the Debye temperature which is a material dependent constant. Although this approach is theoretically sound and describes the heat capacity of a series of elements in their crystalline state very well, it is not suited to

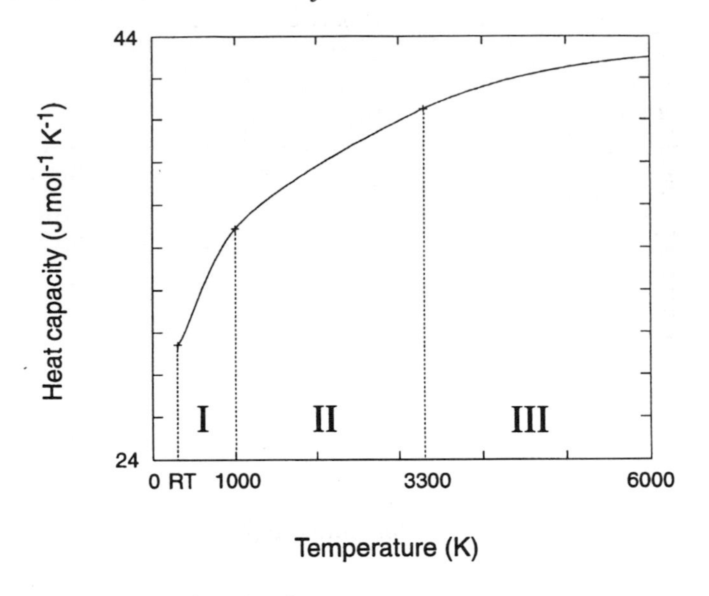

Fig. 2.1 Heat capacity of O_2 (gas).

assess all experimentally known values for solid substances within their respective error limits [69Ger]. Furthermore, for most applications it is not necessary to recur to zero Kelvin as the reference temperature.

Thus a system of thermochemical data has been established on the basis of the standard element reference state (SER). As indicated above, room temperature (298.15 K) and 1 bar of total pressure are introduced as standard conditions, and the enthalpy (H_{298}) of the state of the elements which is stable under these conditions is set to zero by convention. The entropy (S_{298}) is given by its absolute value and the heat capacity at constant pressure, C_p, is described by a polynomial, mostly that introduced by Mayer and Kelley [49Kel]:

$$C_p = c_1 + c_2 T + c_3 T^2 + \frac{c_4}{T^2} \tag{2.12}$$

This approach permits assessment of the thermal properties of most substances within their experimental error limits, as there are sufficiently many adjustable parameters. In some exceptional cases, it may be necessary to split the temperature ranges of the fit to stay within the experimental error limits. As an example, see Figure 2.1 for the heat capacity of O_2(gas).

From the C_p-polynomial and the known values of ΔH_{298} and S_{298} one obtains a Gibbs-energy equation as :

$$G = C_1 + C_2 T + C_3 T \ln T + C_4 T^2 + C_5 T^3 + \frac{C_6}{T} \tag{2.13}$$

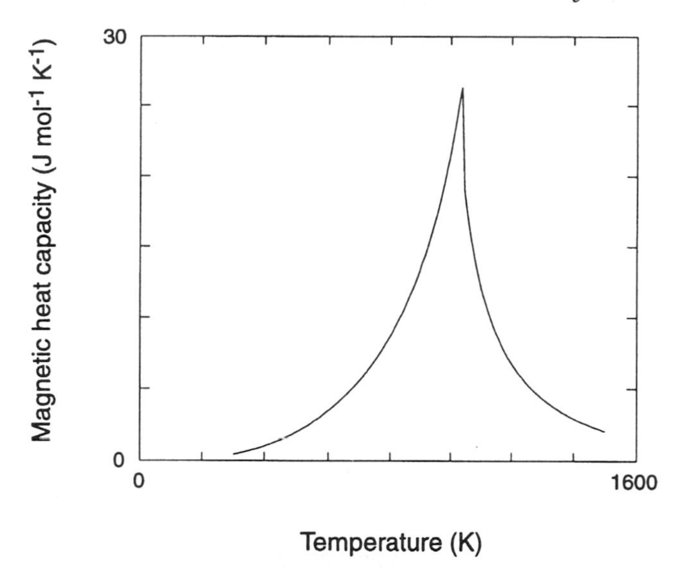

Fig. 2.2 Magnetic contribution to the heat capacity of bcc Fe.

The coefficients C_i are now the data that are to be stored in a Gibbs-energy databank. Note, that the first two coefficients contain contributions from both ΔH_{298} and C_p and S_{298} and C_p respectively, whereas the latter four can be related directly to the four coefficients of the standard C_p equation.

Phase transitions of first order can easily be integrated into this data system, once the temperature and enthalpy of transition and the coefficients of the C_p equation of the phase at higher temperature are known. The G function for the higher range is derived from the integration of enthalpy and entropy now based on the transition temperature instead of room temperature.

The standardised treatment described above has also been used for substances which exhibit magnetic (second order) phase transitions. To be able to handle the anomaly in the heat capacity that arises in such a case (see Figure 2.2), it was thus far customary to split the temperature range around the Curie temperature into several small intervals such that the standard expression for C_p could be used [73 Bar]. This procedure creates an unnecessarily large number of coefficients (e.g. eight times four C_p parameters for Ni) and it also causes numerical difficulties because of the unusually large values of the parameters. SGTE has therefore adopted an approach suggested by Inden [76Ind1,2] which simplifies the situation considerably.

The magnetic contribution is treated separately thus leaving a well behaved curve for the non-magnetic contribution to the heat capacity that can usually be described by one set of standard parameters for the entire temperature range. For the magnetic part of C_p, the critical temperature (T_c, either Curie or Neel temperature), the structure of the phase and the magnetic moment (ß) per atom in this particular lattice are the prerequisites.

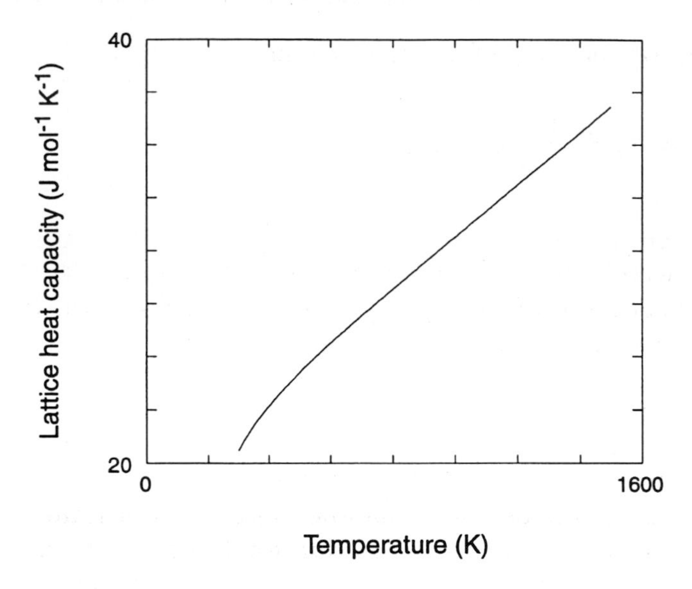

Fig. 2.3 Lattice contribution to the heat capacity of bcc Fe.

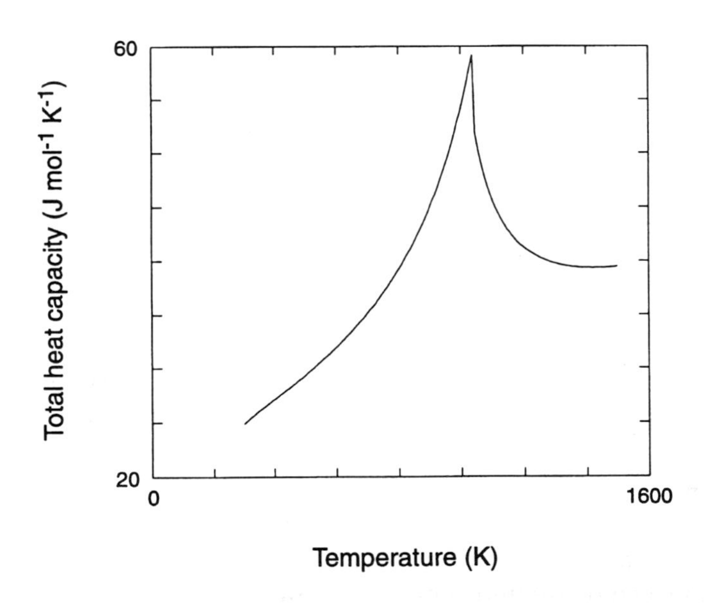

Fig. 2.4 Total heat capacity of bcc Fe.

One obtains for one mole of magnetic element:

$$^{magnetic}G = RTf\left(\frac{T}{T_c}\right)\ln(\beta + 1) \tag{2.14}$$

f is a structure dependent function of temperature. It is different for the ranges above and below the critical temperature. Figures 2.2–2.4 show the two distinct contributions from magnetism and lattice to the heat capacity and the resulting total curve.

A further additive contribution to the Gibbs energy, that is usually ignored because of its negligibly small value, stems from the pressure dependence of the molar volume. However, recent technical developments such as hot isostatic pressing but also more detailed research in geochemical phenomena have created a need to be able to handle this extra contribution. SGTE has adopted the Murnaghan equation [44Mur] for its mathematical description. This equation uses explicit expressions for molar volume at room temperature, V°, its thermal expansion, $\alpha(T)$, the compressibility at 1 bar, $K(T)$, and the pressure derivative of the bulk modulus, n, (Bulk modulus = inverse of compressibility).

$$^{pressure}G = V^0 \exp\left[\int_{298}^{T} \alpha(T)\mathrm{d}T\right] \frac{[1 + nK(T)P]^{\left(1-\frac{1}{n}\right)} - 1}{(n-1)K(T)} \tag{2.15}$$

with $\alpha(T)$ and $K(T)$ polynomials of the temperature.

$$\alpha(T) = A_0 + A_1T + A_2T^2 + A_3T^{-2} \tag{2.16}$$

$$K(T) = K_0 + K_1T + K_2T^2 \tag{2.17}$$

The necessary parameters have so far been assessed for only a few substances, mainly metallic elements and some oxide phases of geological interest. Examples of the resulting P–T phase diagrams, are given in Figures 2.5 and 2.6.

Conclusion

For pure stoichiometric substances the Gibbs energy is only a function of temperature and, if appropriate, total pressure, $G(T,P)$. Different additive contributions can be treated separately. Thus one obtains

$$G_m = {}^{lattice}G + {}^{magnetic}G + {}^{pressure}G \tag{2.18}$$

${}^{lattice}G$ depends upon ΔH_{298}, S_{298} and a C_p-polynomial.
${}^{magnetic}G$ depends upon lattice structure, critical temperature and magnetic moment.
${}^{pressure}G$ depends upon standard molar volume, compressibility and thermal expansion.

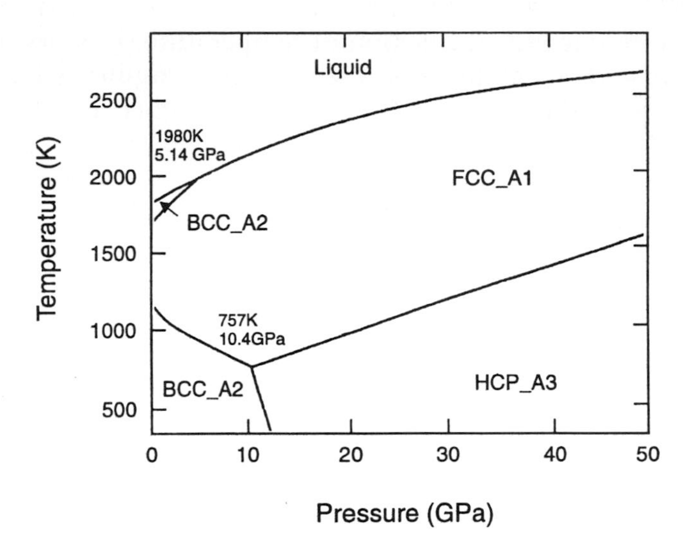

Fig. 2.5 Calculated *P-T* phase diagram for Fe.

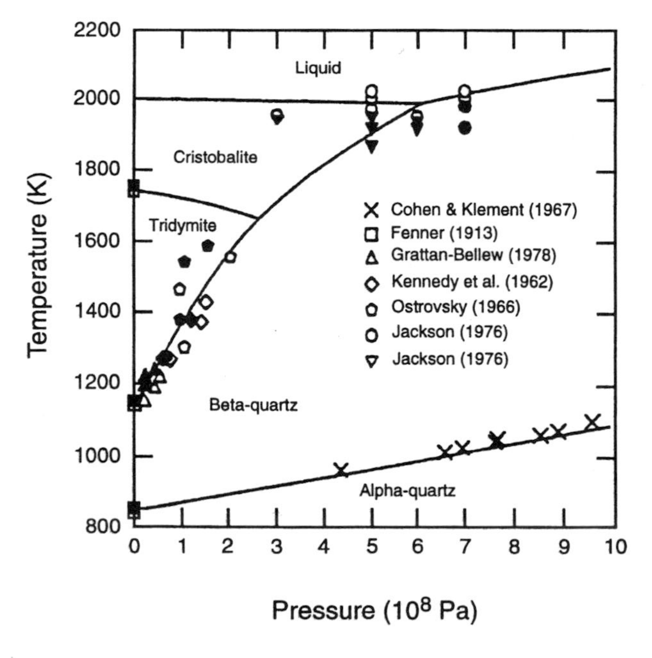

Fig. 2.6 Calculated *P-T* phase diagram for SiO_2.

2.2 Gibbs energies for solution phases

Substitutional solutions

The properties of solutions are usually described relative to the properties of the pure substances in the same structure (φ) and at the same temperature, $\overset{\circ}{G}$:

$$G^{ref} = \sum x_i \, {}^{\circ}G_i^{\varphi} \tag{2.19}$$

where x_i is the mole fraction of constituent i.

In addition one defines an ideal solution by assuming that there is only one further term, that due to the entropy of ideal mixing of atoms. For a substitutional solution all the atoms mix with each other and the term is:

$$G^{id} = RT \sum x_i \ln x_i \tag{2.20}$$

The deviation from this ideal solution model must be described by some mathematical expressions. It is important that these expressions are such that they can be used to estimate the properties of higher order systems from experimental information on the lower order systems. Redlich and Kister [48Red] proposed that one should estimate the properties of a ternary solution from the three component binaries by applying the following type of expression to each binary and evaluate the parameters by fitting to binary experimental information:

$$G_{AB}^{ex} = x_A x_B \sum L_{AB}^{(v)} \left(x_A - x_B\right)^v \tag{2.21}$$

The first term in the Redlich–Kister series is $x_A x_B L_{AB}$ and it is identical to the term in the so called regular solution model. That model can be justified by the Bragg–Williams approach based on random mixing and a consideration of the energies of different kinds of bonds, i.e. between like (AA and BB) and unlike (AB) next nearest neighbours. The L_{AB} parameter should thus be regarded essentially as an energy parameter and it is often called interaction energy. The other $L_{AB}^{(v)}$ parameters may be regarded as describing the composition dependence of L_{AB}.

If there is experimental information on the ternary solution, it can be used to evaluate deviations from the predictions obtained by the Redlich-Kister sum for the binaries. One should first try to describe these deviations with a term $x_A x_B x_C L_{ABC}$ or, if necessary, one should use a power series based on the same principle as the Redlich-Kister series. Information is rarely available from all four ternaries making up a quaternary system and from the quaternary as well. However, if that is the case, one may introduce a quaternary term $x_A\, x_B\, x_C\, x_D\, L_{ABCD}$. The whole expression for the Gibbs energy of a solution phase can thus be given as

$$G_m^{\varphi} = \sum x_i\, {}^0G_i^{\varphi} + RT\sum x_i \ln x_i + G_m^{ex,\varphi} = G_m^{ref} + G_m^{id} + G_m^{ex,\varphi}$$

with

$$G_m^{ex,\varphi} = \sum_i \sum_{<j} x_i x_j \sum_{v=0}^{m_{ij}} L_{ij}^{(v)} (x_i - x_j)^v + \sum_i \sum_{<j} \sum_{<k} x_i x_j x_k L_{ijk}$$
$$+ \sum_i \sum_{<j} \sum_{<k} \sum_{<l} x_i x_j x_k x_l L_{ijkl} + \ldots \qquad (2.23)$$

Partial properties for the substitutional solution

From the expression for the molar Gibbs energy of a substitutional solution, one can calculate the chemical potential for any component element i by the equation:

$$G_i = G_m + \frac{\partial G_m}{\partial x_i} - \sum x_j \frac{\partial G_m}{\partial x_j} \qquad (2.24)$$

where all x_j are treated as independent variables in the derivation.

Also note that the appropriate derivatives of this equation with respect to temperature yield the partial values for enthalpy, entropy and heat capacity.

Sublattice model

Now consider a solid solution where there is more than one kind of lattice site. A simple case, with two sublattices and two elements on each one, would be represented by the formula $(A,B)_a(C,D)_c$, where a and c give the numbers of different sites per formula unit. The simplest model for such a solution would be obtained by assuming random mixing of the atoms within each sublattice. It is then convenient to define mole fractions for each sublattice. They are called site fractions and are denoted y where the superscript s identifies the sublattice. The site fractions are used to define the frame of reference for the Gibbs energy of the solution phase. The Gibbs energy equation for a phase with two sublattices would thus be:

$$G_m^{\varphi} = \sum_i \sum_{<j} y_i^1 y_j^2\, {}^{\circ}G_{ij}^{\varphi} + aRT\sum y_i^1 \ln y_i^1 + cRT\sum y_j^2 \ln y_j^2 \qquad (2.25)$$

This is the so-called 'compound energy' model [81Sun] and $^{\circ}G_{ij}^{\varphi}$ represents

The model can also be used for the case where an element can occupy sites in both sublattice. For instance, the disordered and ordered β–CuZn can be represented by $(Cu,Zn)_1(Cu,Zn)_1$ in which $°G_{Cu,Cu}$ and $°G_{Zn,Zn}$ would represent the Gibbs energy of 2 moles of pure Cu or Zn in the β structure and $°G_{Cu,Zn} = °G_{Zn,Cu}$ would be the main parameter in the description of the transformation to long range order. When $y^1_{Cu} = y^2_{Cu}$ the phase is disordered, when $y^1_{Cu} \neq y^2_{Cu}$ it is ordered.

Partial properties for the sublattice model

The molar Gibbs energy for the (two) sublattice model discussed above is given per mole of formula units of the phase $(A,B)_a(C,D)_c$. The chemical potential of any compound that can be derived by filling the sublattices with one sublattice component each can be calculated from an equation similar to the one for the simple substitutional solution. This time however the site fractions must be used in the partial derivatives. One obtains for example for the compound A_aC_c :

$$G_{A_aC_c} = G_m + \frac{\partial G_m}{\partial y^1_A} + \frac{\partial G_m}{\partial y^1_C} - a\sum y^1_i \frac{\partial G_m}{\partial y^1_i} - c\sum y^2_j \frac{\partial G_m}{\partial y^2_j} \quad (2.26)$$

Interstitial solutions

In an interstitial solution an alloying element does not substitute for a host element but occupies interstitial sites. That kind of solution is described by considering two sublattices, the main (substitutional) sublattice and the interstitial sublattice (Fig. 2.7). It should then be realised that the interstitial sublattice is mainly occupied by vacancies, which must be introduced as a

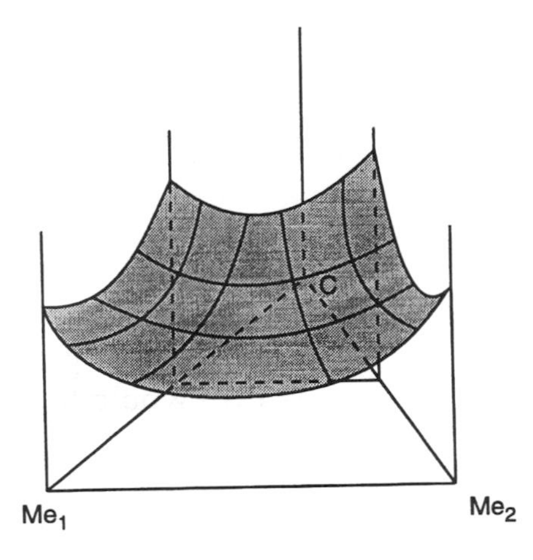

Fig. 2.7 Schematic drawing of the Gibbs energy surface of the fcc_A1 phase with two substitutional metals and interstitial carbon, $(Me_1,Me_2)(Va,C)$.

special element. As an example, the solution of Mn and C in fcc Fe would be represented by the formula $(Fe,Mn)_1(Va,C)_1$. The compound energies, $°G_{Fe,Va}$ and $°G_{Mn,Va}$ would represent the Gibbs energy of the pure fcc Fe and fcc-Mn, but $°G_{Fe,C}$ and $°G_{Mn,C}$ would represent the Gibbs energy of hypothetical carbides, the values of which must be evaluated to fit the experimental information on the solution which does not extent to very high C contents.

From the compound energy model one can primarily calculate the chemical potential for compounds as shown before. For the above example one would thus obtain:

$$\mu_{Fe} = G_{Fe,Va} = G_m + \frac{\partial G_m}{\partial y^1_{Fe}} + \frac{\partial G_m}{\partial y^2_{Va}} - y^1_{Mn}\frac{\partial G_m}{\partial y^1_{Mn}} - y^1_{Fe}\frac{\partial G_m}{\partial y^1_{Fe}} - y^2_{Va}\frac{\partial G_m}{\partial y^2_{Va}} - y^2_{C}\frac{\partial G_m}{\partial y^2_{C}} \quad (2.27)$$

but the chemical potential of the interstitial element would be obtained from a difference:

$$\mu_C = G_{Fe,C} - G_{Fe,Va} = G_{Mn,C} - G_{Mn,Va} = \frac{\partial^2 G_m}{\partial y^2_C} - \frac{\partial^2 G_m}{\partial y^2_{Va}} \quad (2.28)$$

The ideal gas

The gas phase is very often treated by an ideal model, which however takes into account the amounts of all the species that actually exist in the gas. Usually, a minimisation of the Gibbs energy must be performed in order to find the fractions of all species. The Gibbs energy is given for one mole of the final mixture of species and it should be noticed that the number of moles is not known until after the calculation has been performed. The expression for the Gibbs energy is:

$$G_m = \Sigma y_i \, {}^oG_i^{gas} + RT\, y_i \ln y_i + RT \ln P \quad (2.29)$$

and y_i is used here to express the mole fraction among the species, not among the elements. As an example, Figure 2.8 shows the Gibbs energy versus composition for the H–O system at 1 bar. The variable on the composition axis was here taken as $x_{O2} = n_O/(n_H+n_O)$, and, for consistency, the Gibbs energy was not given per mole of actual species but per 2 moles of total H and O, i.e. $2\,n_H + 2\,n_O$. It is interesting to note that the curve has a V shape

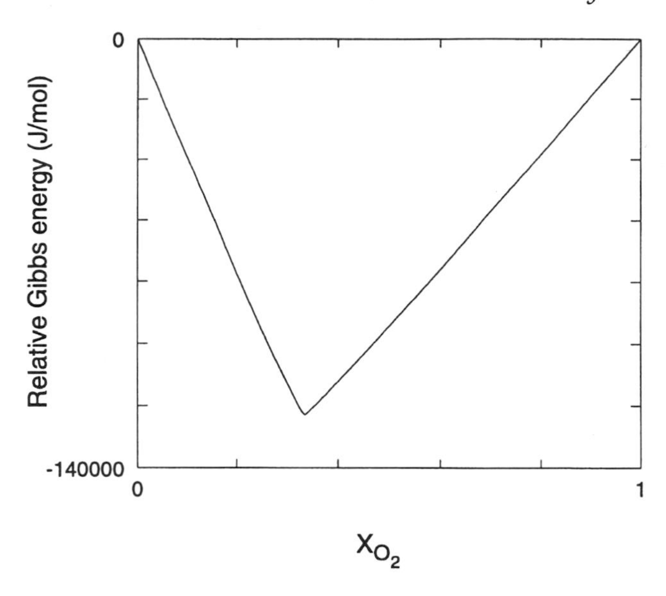

Fig. 2.8 Gibbs energy of the gas phase relative to pure H_2 and O_2, T = 1000K.

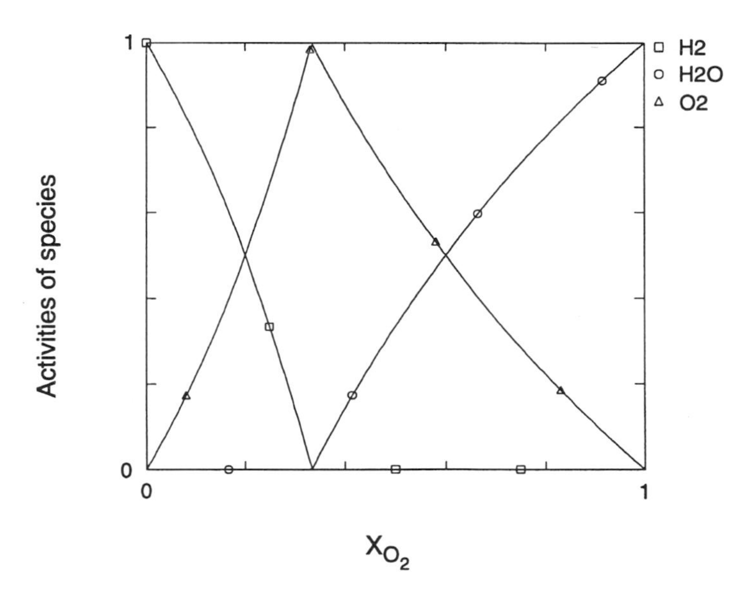

Fig. 2.9 Activities (partial pressures at 1 bar total pressure) of the species in the system H_2 - O_2, T=1000K.

with a very sharp minimum at the position of H_2O. Due to the high stability of H_2O molecules, there are very few other molecules at that composition. To the left of that composition there is hardly any O_2 and to the right hardly any H_2.

This fact is illustrated by Figure 2.9 showing the activities of H_2, H_2O and O_2, where the activities are defined from the chemical potential by

$$\mu_i = {}^{\circ}G_i + RT\ln a_i \tag{2.30}$$

and ${}^{\circ}G_i = {}^{\circ}\mu_i$ is the molar Gibbs energy for pure species i at 1 bar.

Liquid solutions

Many models are used to represent the Gibbs energy of liquid solutions. Assuming that all the atoms mix at random with each other, one can directly apply the substitutional model. Assuming that an element with small atoms dissolves between the other atoms, one can apply the interstitial model. There are many cases where a liquid solution shows a strong deviation from random mixing. Such cases can sometimes be best described by assuming that molecular-like associates form. By treating an associate as a new compound one can again apply the substitutional solution model. For instance, suppose the compound A_2B shows a high stability in the solid state of the A–B system. Then one may assume that A_2B molecules exist in the liquid and one may apply the following expression

$$G_m = y_A{}^{\circ}G_A + y_B{}^{\circ}G_B + y_{A_2B}{}^{\circ}G_{A_2B} + RT\sum y_i \ln y_i + G_m^{ex} \tag{2.31}$$

where y_i is used to define the mole fractions among A,B and A_2B similar to those of a set of gas species. The important parameter in this 'associated solution model' is

$$\Delta{}^{\circ}G_{A_2B} = {}^{\circ}G_{A_2B} - 2\,{}^{\circ}G_A - {}^{\circ}G_B \tag{2.32}$$

The G_m expression for this model is very similar to G_m for ideal gas. When ${}^{\circ}G_{A_2B}$ has a large negative value, then G_m versus composition would be very similar to the H-O case illustrated in Figures 2.8 and 2.9.

Ionic liquids are usually treated by assuming that anions are surrounded by cations and vice versa. According to Temkin [45Tem], this situation can be described by assuming that there is one sublattice for the cations and another one for the anions. The relative number of sites on the two sublattice now depends upon the valencies which can be illustrated by considering the liquid solution of $SrCl_2$ in KCl. It is represented by the formula $(K^+, Sr^{2+})_P(Cl^-)_Q$, Figure 2.10. It is evident that Q/P varies from 1 for pure KCl to 2 for pure $SrCl_2$. In the solid state, there would be two quite different phases, the KCl rich solution $(K^+,Sr^{2+},Va^{\circ})_1(Cl^-)_1$ and the $SrCl_2$ rich solution $(Sr^{2+},K^+)_1(Cl^-,Va^{\circ})_2$.

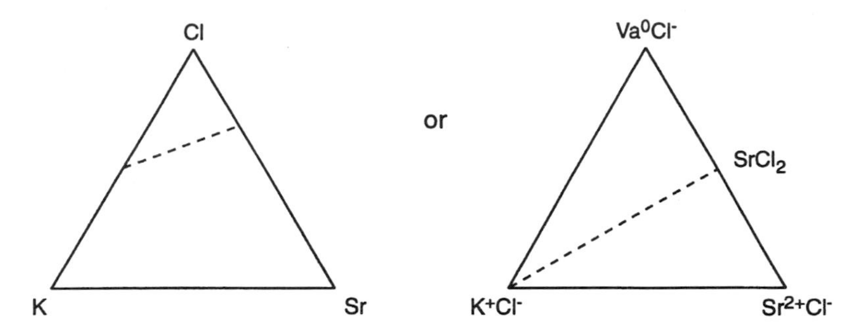

Fig. 2.10 Composition space of electroneutral solution.

It has been proposed [85Hil] that a convenient way of defining the values of P and Q is to identify them with the average value of the ions on the othersublattice. For the liquid solution $(A^+, B^{2+})_P\,(C^-,D^{2-})_Q$ one obtains:

$$-P = -1y^2_C - 2y^2_D \quad \text{and} \quad Q = +1y^1_A + 2y^1_B \tag{2.33}$$

A liquid solution between a metal and a non-metal is sometimes represented by the above ionic two-sublattice model. For the liquid Fe-O solution, the model is defined by the formula:

$$\left(Fe^{2+}, Fe^{3+}\right)_P \left(Va^{q^-}, O^{2^-}, O^0\right)_Q \tag{2.34}$$

The neutral O° atoms have been introduced here in order to allow the solution to extend beyond the O content of Fe_2O_3. Vacancies are introduced in order to allow the model to extend all the way to pure Fe. This is accomplished by assuming that these vacancies are not neutral but have an induced charge equal to minus the average of the charges of the cations, here

$$-q_{Va} = 2y^1_{Fe^{2+}} + 3y^1_{Fe^{3+}} = Q \tag{2.35}$$

The ionic two-sublattice model can also be applied to solutions between metals showing a strong deviation from random mixing. For instance, the liquid phase in the Mg–Sb system, which has a very stable solid phase Mg_3Sb_2, can be described by assuming that Mg is divalent and Sb is trivalent, $(Mg^{2+})_P(Sb^{3-},Va^{q-},Sb^\circ)_Q$. Surprisingly, the result of this model is mathematically identical to the result of the associated solution model if the associates are assumed to contain one atom of the more electronegative element, in this case $Mg_{1.5}Sb_1$.

Molten silicates are occasionally described with the associate solution model using neutral associates like $CaSiO_3$, Ca_2SiO_4 and SiO_2. Sometimes, the ionic two-sublattice model is used with charged associates like SiO_4^{4-} and Ca^{2+} and neutral associates like SiO_2 and $AlO_{1.5}$.

Two more liquid models should be mentioned that have also been used to describe silicate melts. One is the quasichemical model as modified by Pelton and Blander [86Pel]. The other model that has found extensive use is the cell model from Kapoor and Frohberg [71Kap] which was modified by Gaye [84Gay].

Magnetic effects in solution phases

The general equation used for the calculation of the explicit magnetic Gibbs energy contribution has already been given for pure magnetic substances. It contains the critical temperature T_c and the magnetic moment β. For solution phases both are functions of the composition of the phase. This is the only difference between pure magnetic substances and solution phases. The critical temperature separates the range of either ferro-magnetism or antiferro-magnetism from the range of paramagnetism; that is why it is termed critical temperature instead of Curie- or Neel-temperature. However, as ferro- and antiferro-magnetism are mutually exclusive it is possible to treat them in the same formalism. This is illustrated qualitatively in Figure 2.11 for the case of the bcc_A2 phase in the Fe-Cr system.

The solid line shows the 'real' critical temperature as a function of x_{Cr}. Note that T_c must always be greater than or equal to zero! The zero-point separates the ferro- from the antiferro-magnetic range. However, formally it is possible to calculate one continuous and steady curve by introducing 'negative' critical temperatures for the antiferro-magnetic range. These have to be taken as their absolute values to get the real Neel temperature.

However, for the fcc_A1 phase the situation is slightly more complex. The change from ferro- to antiferro-magnetism depends upon the number of nearest neighbours. For the bcc_A2 phase, this has no effect since the coordination number does not change upon change of the spin direction (ferro —> antiferro), because bcc_A2 is a symmetrical structure. For fcc_A1, the situation is different since only 1/3 of the changed spins will contribute in the antiferro-magnetic range, the remaining 2/3 cancel each other. Therefore, a correction by the factor 1/3 before mirror imaging into the positive range is required. The structure dependent factor is thus defined to be 1 for bcc-structures and 1/3 for fcc-structures (and also hcp-structures).

For a solution between a ferromagnetic and an antiferromagnetic element, T_c and β must go through zero at the same composition in order for the model to yield reasonable results. In general, the composition dependence of T_c and β may be described with the Redlich–Kister type of polynomial series.

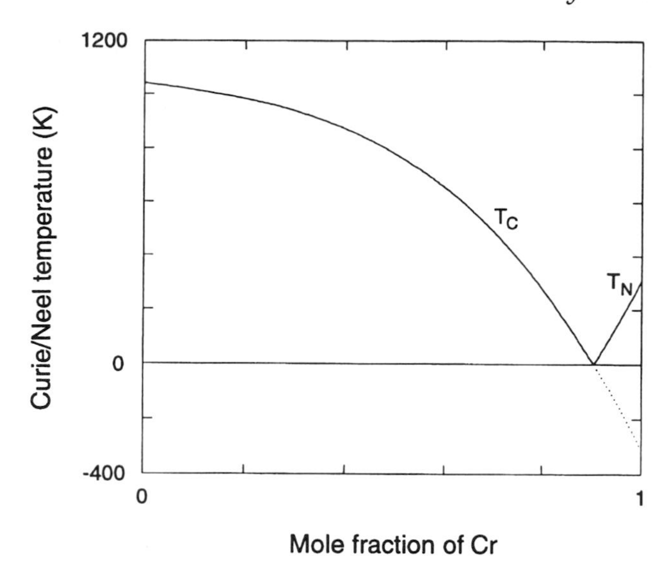

Fig. 2.11 Curie/Neel temperature for 1 mol of the bcc_A2 phase versus composition in the system Fe–Cr, SGTE data used.

For more details on modelling, the reader is referred to scientific journals such as Metallurgical Transactions, Chemica Scripta, Calphad, Journal of Physics and Chemistry of Solids, Geochimica et Cosmochimica Acta and Physics and Chemistry of Minerals.

References

07Ein A.Einstein: *Ann. Physik* **22**, 1907, 180.

12Deb P.Debye: *Ann. Physik* **39**, 1912, 787.

13Fen C.L.Fenner: *Am. J. Sci.*, **36**, 1913, 331-384

44Mur F.D.Murnaghan: *Proc. Natl. Acad. Sci. (USA)* **30**, 1944, 244.

45Tem M.Temkin: *Acta Phys. Chim. USSR* **20**, 1945, 411.

48Red O.Redlich and A.T.Kister: *Ind. Eng. Chem.* **40**, 1948, 345-348.

49Kel K.K.Kelley: *US Bur. Mines, Bull. No. 476*, 1949.

62Ken G.C.Kennedy, G.J.Wasserburg, H.C.Herd and R.C.Newton: *Am. J. Sci.*, **260**, 1962, 501-521

66Ost I.A.Ostrovsky: *Geol. J.*, **5**, 1966, 127-134

67Coh L.H.Cohen and W.Klement Jr: *J. Geophys. Res.*, **72**, 1967, 4245-4251

69Ger C.Gerthsen and H.O.Kneser: *Physik*, Springer Verlag Berlin, Heidelberg, New York, 1969.

71Kap M.L.Kapoor and G.M.Frohberg: *Proc. Symp. Chemical Metallurgy of Iron and Steel*, Sheffield, 1971, 17-22.

73Bar	I. BARIN and O. KNACHE: Springer-Verlag, Berlin, Heidelberg, und Verlag Stahleisen, Düsseldorf, 1973
76Jac	L.JACKSON: *Phys. Earth Planet. Inter.*, **J3**, 1976, 218-231
76Ind1	G.INDEN: *Proc. of Calphad* **V**, Düsseldorf, 1976, III.4, 1-13.
76Ind2	G.INDEN: *Proc. of Calphad* **V**, Düsseldorf, 1976, IV.1, 1-33
78Gra	P.E.GRATTAN-BELLEW: *Expl. Miner*, **11**, 1978, 128-139.
81Sun	B.SUNDMAN and J.ÅGREN: *J. Phys. Chem. Solids* **42**, 1981, 297-301.
84Gay	H.GAYE and J.WELFRINGER: *Proc. 2nd Internat. Symp. Metal. Slags and Fluxes,* H.A.Fine and D.R.Gaskell eds, Metall. Soc. AIME, New York, 1984, 357-375.
85Hil	M.HILLERT, ,B.JANSSON, B.SUNDMAN and J.ÅGREN: *Met. Trans. A* **16A**, 1985, 261-266.
86Pel	A.D.PELTON and M.BLANDER: *Met. Trans. B* **17B**, 1986, 805-815.

3 Graphical Representations of Equilibria

KLAUS HACK

The general relationships between the thermochemical potentials temperature, pressure, chemical potential etc., their conjugate extensive properties entropy, volume, number of moles etc. and the three general types of phase diagrams have been discussed by Pelton and Schmalzried [73Pel]. However, the particular diagrams which result from computer calculations for two dimensional sections of multicomponent systems, i.e. systems with more than three components, are still rather unfamiliar. A second type of diagram, which results from a one dimensional section through a system, has also been made possible by computer calculations. Both types, general two-dimensional phase maps and socalled property diagrams, will be discussed below.

Zero phase fraction lines and two-dimensional phase maps

Masing [44Mas] stated in 1944 that 'a state space can ordinarily be bounded by another state space only if the number of phases in the second space is one less or one greater than that in the first space considered'. In the 1950s and 60s Palatnik and Landau [64Pal] discussed in a series of publications the dimensionality of phase boundaries from which comes the law of adjoining phase regions:

$$R^* = R - D^- - D^+ \geq 0 \tag{3.1}$$

In this equation, R^* is the dimensionality of the phase boundary, R is thedimension of the phase diagram (this may also be a section, isothermal or isopleth), and D^- and D^+ represent the number of disappearing and appearing phases when crossing the phase boundary. For two dimensional diagrams, as discussed here, R=2. For the boundaries one will, therefore, obtain either a dimensionality of 1, i.e. a line, or zero, i.e. a point, because $R^* \leq R–1$, as was shown by Prince [66Pri]. These are all topological elements possible in a two-dimensional graph.

Recently Moral and co-workers [84Mor,84Bra,86Gup] have published a series of papers in which they introduced the concept of Zero-Phase-Fraction (ZPF) lines. This adds a very useful aspect to the lines and points mentioned above so far. It is illustrated by the following series of figures (Figs 3.1-3.3). Consider an isothermal section of a ternary phase diagram, for example for the system A-B-C with the three terminal solution phase alpha, beta and gamma. If one draws such a diagram, one would usually start drawing the lines around the phase regions of the pure phases, add the tie-triangle around the three phase region and possibly the tie-lines in the two phase regions.

From the tie-lines,one obtains addidional information: the amount of the phase present in the two or three phase equilibrium. This is given by the lever rule and can be read directly from a figure such as Fig. 3.1. By drawing the lines for 50 and 0 mole percent, Figure 3.2 is obtained.

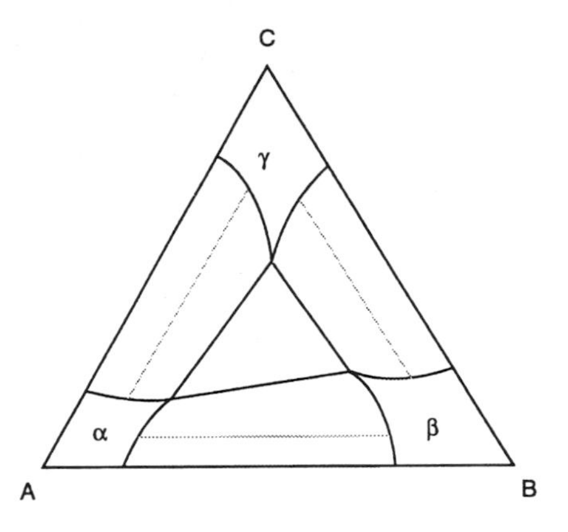

Fig. 3.1 Conventional representation of an isothermal section (extensive property diagram).

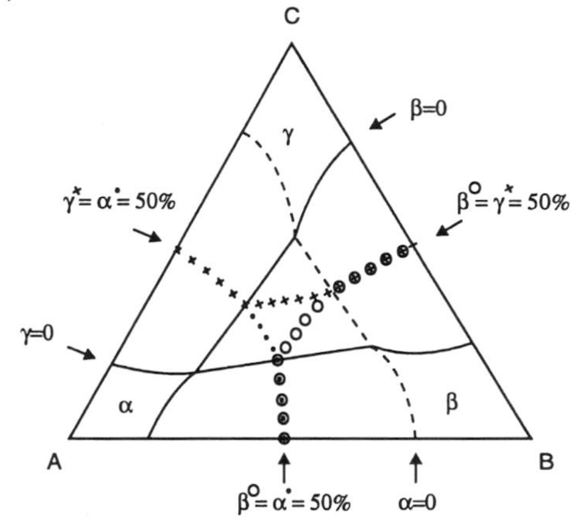

Fig. 3.2 The lines of 50 and 0 mole percent of the respective phases.

The most interesting lines are those for the zero phase amount (or fraction). These lines show for each phase a composition range in which it exists, whether alone or together with others. Equally, the lines, when seen from the other side, indicate that range in the diagram in which a particular phase does not exist. For the alpha-phase, for example, the zero phase fraction line runs as shown in Fig. 3.3.

This aspect was not explicitly included in Masing's or Palatnik and Landau's considerations or Princes's interpretation of them, but it helps considerably to understand a diagram such a Fig 3.6, as the 'inside-outside' property (see Fig. 3.4) of a ZPF yields a very simple pairwise relationship of the lines and phases in a diagram. The zero phase fraction line is as shown in Figure 3.3.

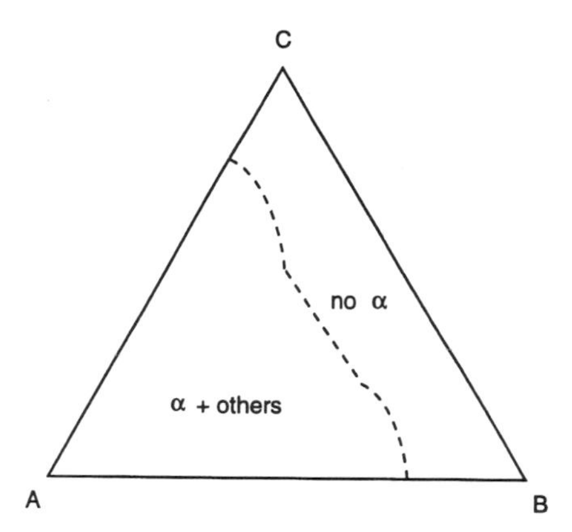

Fig. 3.3 The ZPF line for the α phase.

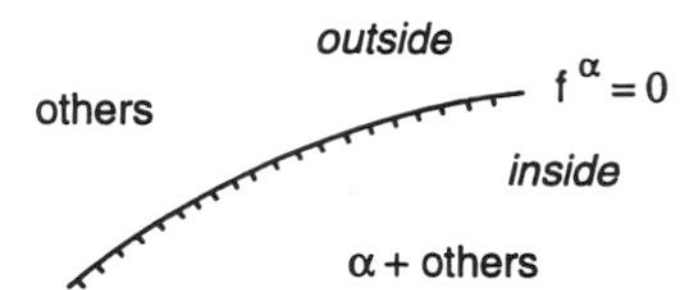

Fig. 3.4 The inside-outside property of a ZPF line.

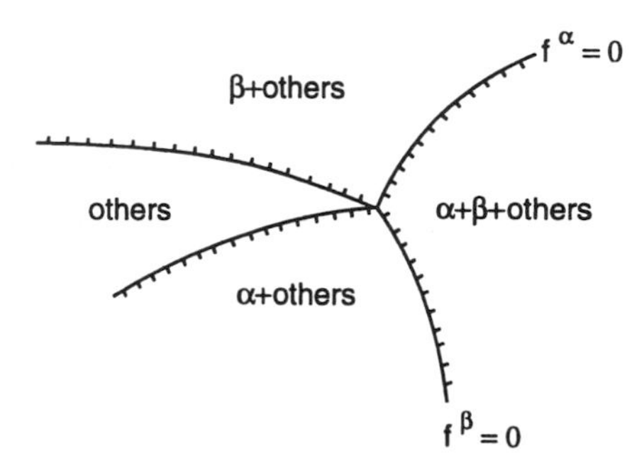

Fig. 3.5 The intersection of the ZPF lines.

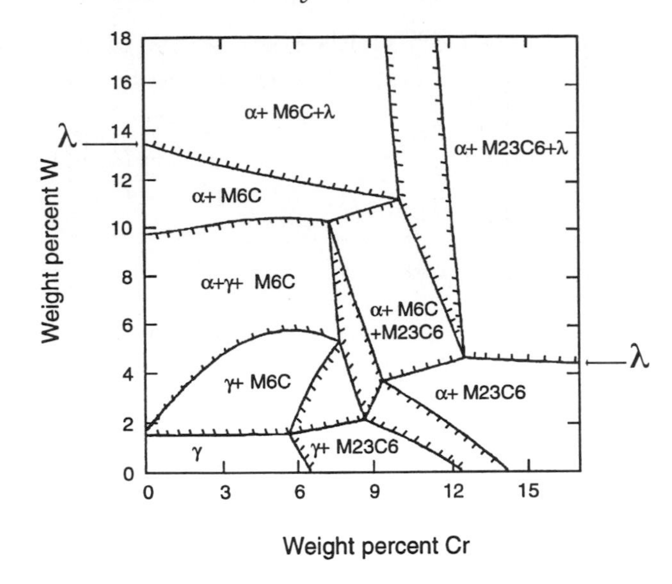

Fig. 3.6 Marked ZPF lines in the Fe–W–Cr–C system with C = 0.2 wt% and T=1123 K. The inside of each ZPF line is marked. Please try to work out the phase related to each line. The line for the λ-phase is marked.

To follow the line of a particular phase through a point of intersection with another line means to go 'straight' through the crossing. At the crossing the law of adjoining phase regions holds for the four lines and the point of intersection. This is shown in Fig. 3.5.

Further features of ZPF-lines are that they begin and end on the diagram axes or else form a closed loop within the diagram. The first of these features can easily be understood from Fig. 3.3.

Knowing the features of ZPF-lines as discussed above, Figure 3.6, an isothermal section of a four component system, can now easily be interpreted considering all lines as zero-phase fraction lines.

Special cases

The more components and phases a system contains, the more likely it is that the rules outlined above are sufficient to understand calculated two-dimensional phase diagrams. However, there are some exceptions that occur and which have to be discussed to prevent misunderstandings.

The general case is that four lines meet at a point and the law of adjoining phase regions holds. Both these rules seem to be violated in a simple binary eutectic system as shown in Figure 3.7.

At points *A* and *C* only three lines meet, and at point *B*, going from the liquid one-phase region into the solid two-phase region, the number of phases changes from 1 to 2 instead of 3. The reason for this behaviour is that the three phase field *A–B–C* is what Prince calls a degenerate phase region.

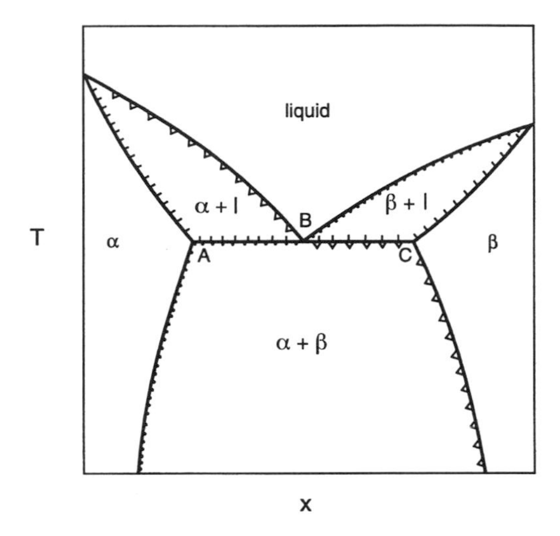

Fig. 3.7 Simple binary eutectic system, ZPF lines Δ = α, – = liquid, • = β.

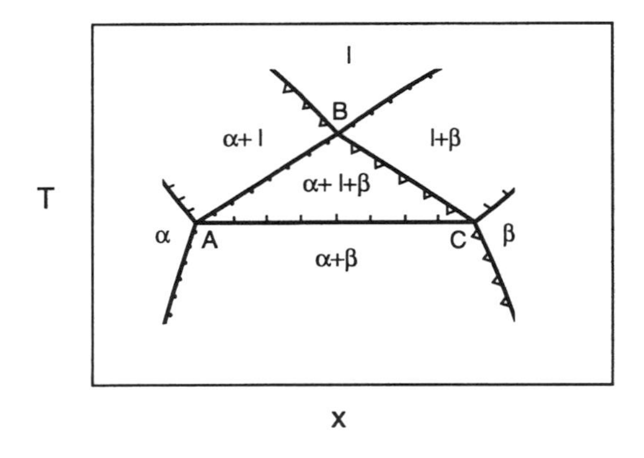

Fig. 3.8 Exaggerated plot of α–liquid–β three phase region, ZPF lines Δ = α, – = liquid, • = β. Note that addition of a third component will naturally yield this topology.

Instead of one line, there are in fact three, (A–B, B–C and A–C), separating the three-phase region alpha-liquid-beta from the respective two-phase regions. An exaggerated drawing with a finite angle between the lines shows this very clearly, Figure 3.8.

Now there are only points at which four lines meet and at point B, going down in temperature from the one-phase region, three phases are formed in accordance with the law of adjoining phase regions.

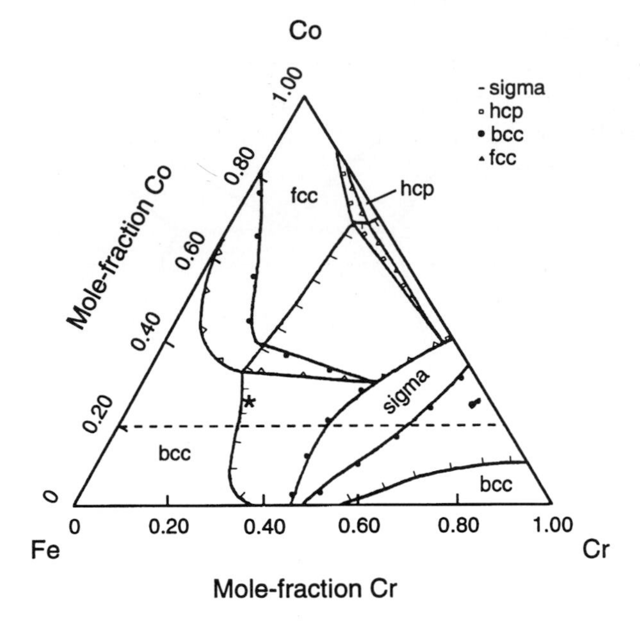

Fig. 3.9 The system Co–Cr–Fe for $T = 1073K$, * = Fe 50, Co 25, Cr 25 At%, Zero-phase fraction lines marked (Dashed line and * see Figs. 3.10 and 3.11).

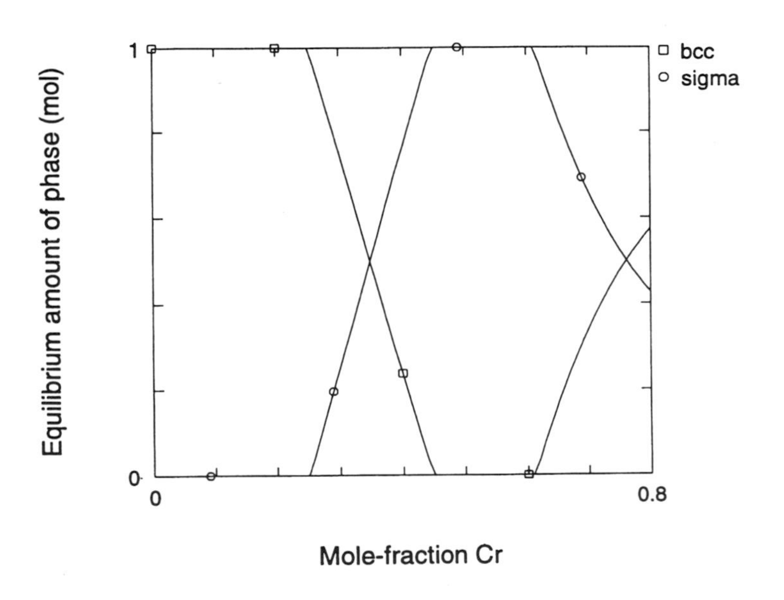

Fig. 3.10 One dimensional phase map along the dashed line in Fig. 3.9.

Property diagrams

This type of diagram is directly related to many practical applications of thermochemistry, e.g. the experimental investigation of an alloy sample using a differential thermal analysis or the heat treatment of a particular alloy. Following is an example of one-dimensional phase maps applied to the Co–Cr–Fe system.

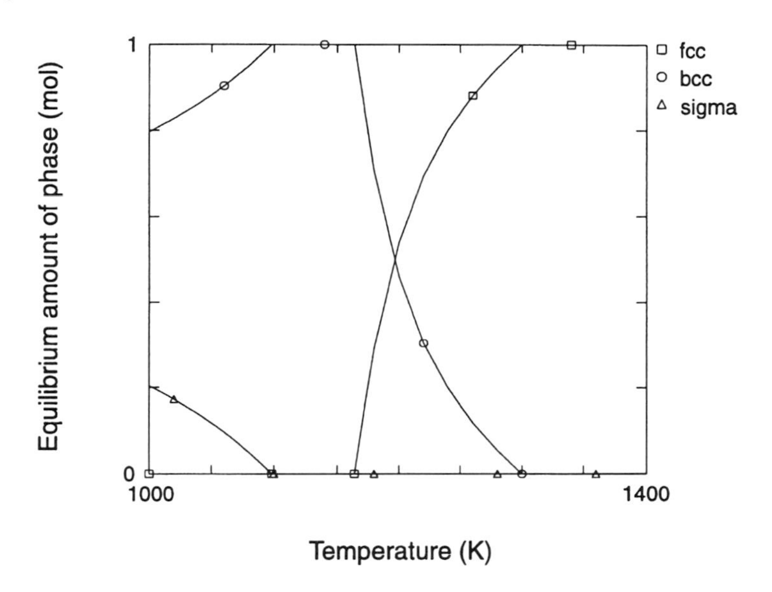

Fig. 3.11 One dimensional phase map through point (*) in Fig. 3.9.

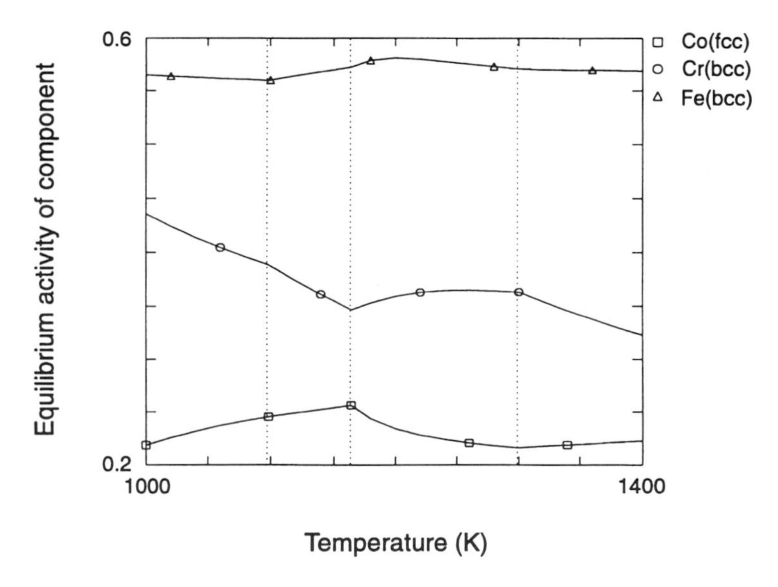

Fig. 3.12 Component activities according to Fig. 3.11.

For a first orientation, a two-dimensional phase map is used, here an isothermal section of the system for T=1073 K (and P=1 bar), Fig. 3.9. One dimensional phase maps can easily be related to this diagram by selecting an extensive property or a potential axis which runs through it.

The first case, (Figure 3.10), shows properties for constant temperature along the line of composition marked in Figure 3.9. The activities of the system components have been chosen as the dependent quantities. Note that the kinks in the curves reflect the crossing of phase boundaries in Figure 3.9. The second case (Figure 3.11) shows properties along a line which runs perpendicularly through the phase diagram in Fig. 3.9. For the composition Fe50Co25Cr25 at.% (marked by an asterisk in Figure 3.9) the temperature has been varied through a certain interval, 1000K $< T <$ 1400K. The amounts of the equilibrium phases have been chosen as dependent variables. In terms of the completeness of the result of complex equilibrium calculations, it is worth mentioning here that using the same results it is also possible to draw Fig. 3.12. This figure shows the equilibrium activities of all elementary components of the system as a function of temperature for the given composition.

References

44Mas	G.Masing: *Ternary Systems,* Reinhold, New York, 1944.
64Pal	L.S.Palatnik and A.I.Landau: *Phase Equilibria in Multicomponent Systems,* Holt, Rinehart and Winston, New York, 1964.
66Pri	A.Prince: *Alloy Phase Equilibria,* Elsevier, Amsterdam, London, New York, 1966.
73Pel	A.D. Pelton and H.Schmalzried: *Met. Trans* **4**, 1973, 1395
84Bra	T.R.Bramblett and J.E.Morral: *Bull. Alloy Phase Diagrams* **5**, 1984, 433-436.
84Mor	J.E.Morral: *Scripta Metallurgica* **18**, 1984, 407-410.
86Gup	H.Gupta, J.E.Morral and H.Nowotny: *Scripta Metallurgica* **20**, 1986, 889-894.

4 Summarising Mathematical Relationships between Gibbs Energy and other Thermodynamic Information

KLAUS HACK

In the preceding chapters it was demonstrated how the different mathematical relationships of thermodynamics can be applied to a great variety of different tasks, from the simple calculation of thermodynamic properties of pure substances to that of a phase diagram of complex multi-component multi-phase systems. A schematic summary is shown in Figure 4.1 of the mathematical tools that relate the Gibbs energy of a system to the various other thermodynamic information.

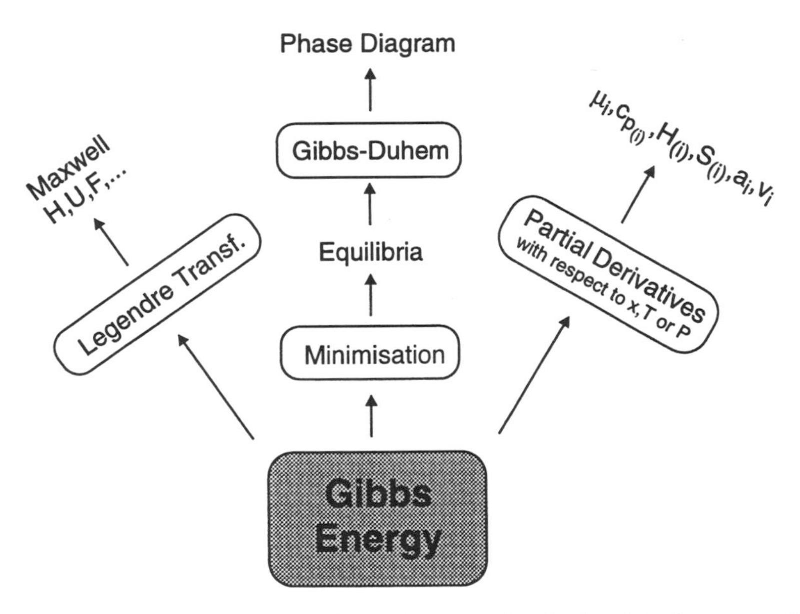

Fig. 4.1 Schematic summary of mathematical relationship between Gibbs Energy and other thermodynamic parameters.

- Using Gibbs Energy (G) as a function of T, P and n_i, Legendre transformations yield all other thermodynamic *potentials* with respect to their natural variables. This was already pointed out by Gibbs but is now better known as the Maxwell relationship.

- Using the appropriate partial derivatives of G with respect to T, P and n_i, all integral and partial values of Gibbs energy, entropy and heat content of a phase can be calculated. This can be done by programming the derivatives explicitly for all Gibbs energy models based on explicit mathematical expressions of T, P and the concentration variables. However, it should be noted that some models, such as the cell model or the quasi-chemical model for ionic liquids, require a phase internal minimisation with respect to composition and numerical differentiation (with all inherent disadvantages).

- Based on G, minimisation leads to equilibrium states. In order to obtain a complete description of the respective system, including both chemical potentials and equilibrium amounts, the method of 'extent of reaction' can be applied for equilibria involving simple stoichiometric reactions. A method using Lagrangian undetermined multipliers is usually applied for calculations involving multi-component, multi-phase systems. It should be noted that both methods require numerical procedures, i.e. the use of a computer program.

- From single equilibrium calculations, the phase mapping techniques based on the Gibbs-Duhem equation lead to phase diagrams of three different but interrelated kinds (potential, mixed, and extensive property diagrams). To obtain quantitative relationships in multi-component systems, two-dimensional sections can be generated, again using appropriate computer programs.

Part II: Applications in Materials Science and Processes

Introduction

Thermodynamic analysis can be used as a tool in many, and extremely diverse fields; for example:

- for handling problems involving extremes of temperature - from room temperature, e.g. salt solutions in water [84Har], to several hundred degrees, e.g. fuel behaviour in combustion engines [89Mor], or up to several thousand degrees, e.g. technical plasmas [82Lyh];

- for improving low technology applications, such as the aluminothermic welding of rails [87Hac], as well as for developing high technology processes, such as liquid phase epitaxy in micro-chip manufacture [91Mad];

- for understanding generally acceptable technologies, e.g. processes for producing electric light bulbs [92Muc], as well as for far more controversial ones, e.g. safety analysis in nuclear plants [83Pot];

- for explaining geophysical phenomena at the extremes of our solar system, e.g. the core of the earth and the solar nebulae [84Poi, 86Sax].

An important reason for employing thermodynamic methods is that they offer a means of systematising and modelling the behaviour of interacting phases in complex systems, allowing the phase relations existing in multicomponent, multiphase systems to be described, as well as in some cases, their subsystems. It is seldom necessary to have thermochemical data available for systems of more than three or four components in order to make reasonable calculations for higher order systems. Consequently thermodynamic calculations can also constitute a powerful method for guiding experiment. The reverse is also true, a thermodynamic analysis is best undertaken in the context of a knowledge of the materials science of the problem. Hence a combined approach using thermodynamic calculations coupled with experimental measurement is frequently highly effective.

The areas in which computational thermochemistry finds application are constantly increasing. Therefore, the examples listed in the table below are only intended to provide an insight into the extremely diverse nature of some of those areas in which computational thermochemistry has been applied in inorganic materials and process research and development. An equally long list could be put together for organic materials.

Some areas of application of computer assisted thermochemistry

Sintering
Roasting
Purification of metals
Waste treatment
Heat-treatment cycles
Alloy design
Precipitation in alloys
Dispersion strengthening
Hard metal compound phases
Combustion
Nitrates in water
Steels
Tungsten halogenides
Cement manufacture
Dioxine emissions
Glasses
Aluminium electrolysis
SiC in silver alloy composites
Casting processes
Incineration
Electron beam melting
Ore reduction
Corrosion
Nozzle clogging during casting
Extractive metallurgy
Electronic materials
High T_c superconductors
Ceramics
Slags
Molten salt batteries
VLE/LLE of organic solutions
Geochemistry/Planetology

Problem analysis

A number of general considerations are relevant when applying computer assisted thermochemistry. These can be summarised in the following description of the three stages customarily applied during the calculation procedure.

I Firstly, a definition of the problem is made in terms of global conditions, such as temperature, total pressure, overall composition of the system, possible constraints such as fixed activities and/or partial pressures of certain system components, etc. A minimum list of elements necessary to treat the problem is fixed. In addition, the possible phases are specified and, where applicable, a suitable reference substance is chosen.

II The second step consists of checking the available thermochemical data for the elements, and particularly for the phases and their components. The problem can then be transformed into a set of variables suitable for a computer calculation, and the most suitable type of program for the case is selected. Sometimes the thermochemical data are incomplete and it may be necessary to make estimations for missing data. In the meantime, an extensive range of methods is also available for this purpose in computerised form [93Kub]. In other cases, raw experimental data may be available from the literature that can be assessed to prepare a consistent data-set.

III After completing the calculation, the third step in the procedure is to interpret the results in the context of the chemistry of the problem. This analysis enables major influences to be determined, and in many cases optimum values to be calculated.

It is useful to keep these considerations in mind when reading the different case studies described in Part II.

References

82Lyh W.LYHS and ,H.WILHELMI: *Arch. Eisenhüttenwesen* **53**, 1982, 49–54.

83Pot P.E.POTTER and M.H.RAND: *Calphad* **7**, 1983, 165-174.

84Poi J.P.POIRIER: *Calphad XIII*, Villar de Lans, 1984.

84Har C.E.HARVIE, N.MOELLER AND J.H.WEARE: *Geochem. Cosmochem. Acta* **48**, 1984, 723.

86Sax S.K.SAXENA AND ,G.ERIKSSON: *Adv. Phys. Geochem.* **6**, 1986, 30-105.

87Hac K.HACK: *Case studies for training course in Theoretical Metallurgy*, LTH - RWTH Aachen, 1987.

89Mor A.MORGAN and K.MEINTJES: *Combust. Sci. and Tech.* **68**, 1989, 35-48.

91Mad R.MADAR and C.BERNARD: *Appl. Surface Science* **53**, 1991, 1-10.

92Muc S.A.MUCKLEJOHN and A.T.DINSDALE: *6th Int. Symp. on the Science and Technology of Light Sources*, Budapest, 1992, 149-151.

93Kub O.KUBASCHEWSKI, C.B.ALCOCK and P.J.SPENCER: *Materials Thermochemistry*, Pergamon, Oxford, 1993.

5 Hot Salt Corrosion of Superalloys

TOM I. BARRY[1] AND ALAN T. DINSDALE[2]

Introduction

Under normal circumstances a nickel based superalloy is protected from corrosive attack by an oxide layer on the surface which is typically Cr_2O_3. However in marine environments turbine blades made from superalloys are particularly susceptible to corrosive attack by hot or molten salts. Under these conditions NaCl can be swept into the turbine in the air stream to react with any sulphur containing combustion products in the fuel to form sodium sulphate which condenses onto the oxide layer. This salt droplet may dissolve the protective layer and expose the alloy to corrosive attack from the oxidising and sulphidising atmosphere. In this paper the range of conditions which might lead to this so-called 'hot salt corrosion' are explored in an attempt to provide a basis for understanding the detailed mechanisms involved.

The driving force for corrosion is thermodynamic and hence it is attractive to analyse corrosion by reference to thermodynamic criteria. On the other hand corrosion involves a series of processes which are linked together in a complex way. For the case of hot salt corrosion the mechanism by which chromium is removed from the surface of the turbine blade is particularly complex and may involve some sort of fluxing whereby the Cr_2O_3 is reprecipitated from the salt phase at its boundary with the gas phase. Traditionally corrosion problems have been analysed thermodynamically in terms of so-called phase stability diagrams. Such diagrams are conceptually simple but are really limited to systems where the material is a pure element. In view of the progress made in calculating phase equilibria involving many components and a wide range of phases it is appealing to analyse the corrosion phenomenon as a multicomponent system.

Data used for the calculations

In order to obtain some thermodynamic insight into this process, data for the C-H-O-N-Cl-S-Cr-Na system from the SGTE pure substance database were augmented by data assessed for the liquid phase and various solid

1 Amethyst Systems, Marlingdene Close, Hampton, Middlesex, TW12 3BJ, UK
2 Centre for Materials Measurment and Technology, National Physical Laboratory, Teddington, Middlesex TW11 0LW, UK

salt phases thought to be of importance. Of the phases involved in hot corrosion only the gas phase approximates closely to thermodynamically ideal behaviour. The phases present in the alloy, corrosion products and adherent salts are all non-ideal solutions.

The chemical properties of the deposited salt are usually analysed in terms of its oxide activity. However, since NaOH is much more stable than Na_2O in the presence of the water in the gas stream it was decided to analyse the data required for calculations in terms of the quaternary system

$$NaCl–NaOH–Na_2CrO_4–Na_2SO_4$$

Five condensed solution phases have been considered in this treatment, the liquid, the orthorhombic and hexagonal structure of Na_2SO_4, the halite structure of NaCl and the monoclinic (high temperature) structure of NaOH. Mixing of all components was considered for the liquid but only pairs of components in the crystalline phases, as shown in Table 5.1.

Table 5.1 List of condensed salt phases considered and their constituents.

liquid	NaCl, NaOH, Na_2CrO_4, Na_2SO_4
halite	NaCl, NaOH
monoclinic NaOH	NaCl, NaOH
hexagonal Na_2SO_4	Na_2CrO_4, Na_2SO_4
orthorhombic Na_2SO_4	Na_2CrO_4, Na_2SO_4

None of these salt phases required a large number of terms to model the solution behaviour and some approached ideality. The assessments for the systems NaCl–Na_2SO_4, NaCl–Na_2CrO_4 and Na_2SO_4–Na_2CrO_4 are based on the work of Liang, Lin and Pelton [80Lia 1,2] with some modifications. Different data are used for the pure components and solution data are added for the orthorhombic sodium sulphate phase. The NaOH–Na_2SO_4 system has been assessed by Bale and Pelton [82Bal] but, since in that particular case equivalent rather than mole fractions were used, a complete reassessment was necessary. New assessments have also been made for the NaCl–NaOH and the NaOH–Na_2CrO_4 systems [87Bar]. The most important salt binary system is the Na_2SO_4–Na_2CrO_4 system in which the components are completely soluble in each other in both the liquid phase and in the hexagonal and orthorhombic Na_2SO_4 based solid solution phases. Figure 5.1 shows the calculated phase diagram for this system.

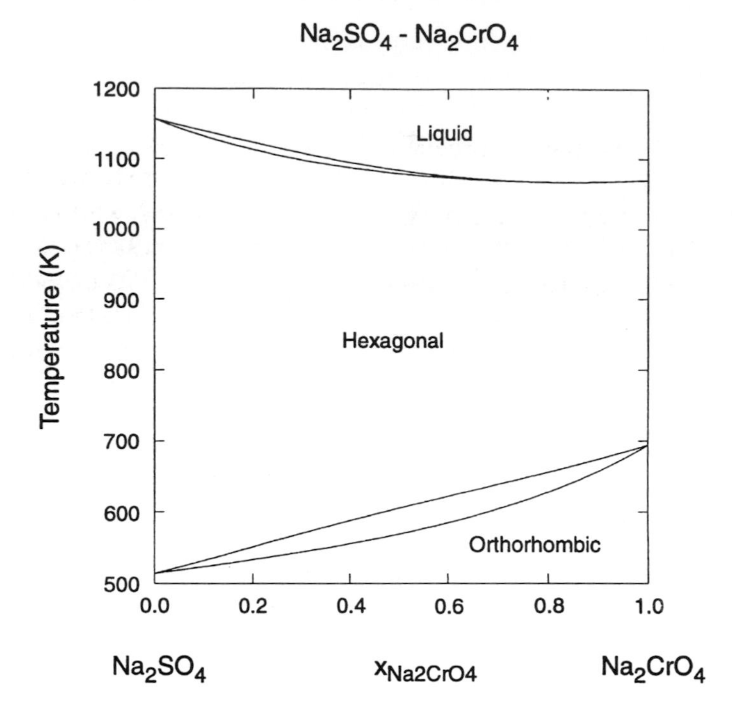

Fig. 5.1 Calculated phase diagram for the Na_2SO_4-Na_2CrO_4 binary system.

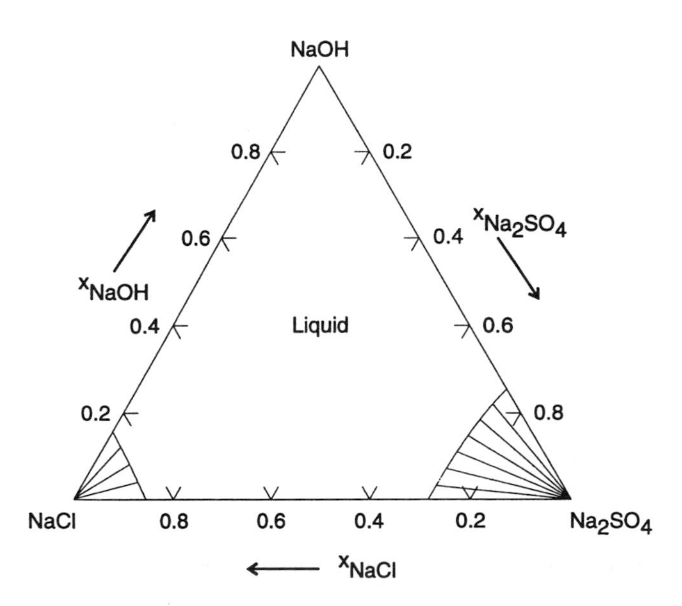

Fig. 5.2a Phase diagram for 1023.15 K for ternarysystem NaCl-NaOH-Na_2SO_4

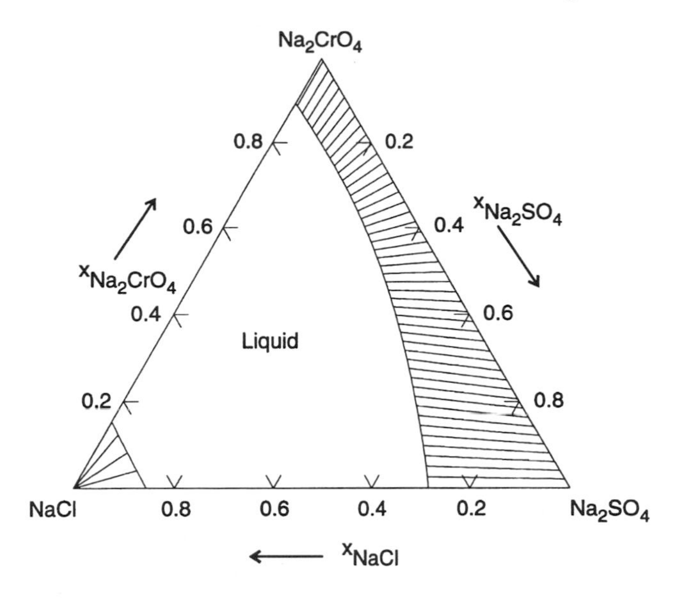

Fig. 5.2b Phase diagram for 1023.15 K for the ternary system NaCl-Na_2SO_4-Na_2CrO_4.

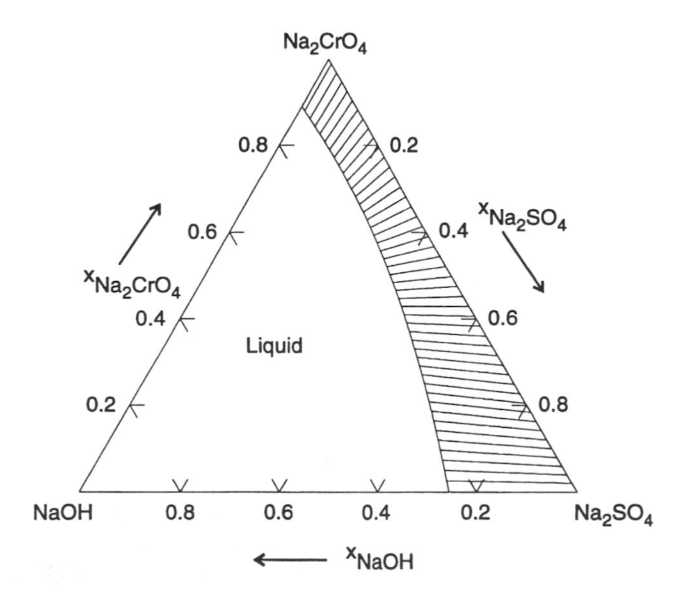

Fig. 5.2c Phase diagram for 1023.15 K for the ternary system NaOH-Na_2SO_4-Na_2CrO_4.

Figures 5.2 a-c show the diagrams for three of the four ternary systems at 1023.15 K calculated from the critically assessed binary data.

The gas-salt equilibrium

The following calculations relate to the corrosion in a gas turbine operating at a fuel-air ratio of 50:1 by weight, a pressure of 15 bar (1.5 MPa) and at a temperature of 750°C. The fuel composition was taken to have the approximate formula $CH_{1.8}$. With these assumptions the main combustion products have the following partial pressures in bars

CO_2: 0.61; H_2O: 0.55; N_2: 11.6; O_2: 2.216.

On this basis 1% sulphur in the fuel would cause the sum of the pressures of SO_2 plus SO_3 to be 0.0026 bar. The presence of 1 ppm of sodium chloride by weight of gas would result in a pressure of HCl of 0.0046 bar. The sodium will accumulate mainly as the sulphate on any surfaces. Hence the amount of sodium that can be used in the calculation is arbitrary. On the other hand, for calculations relating to the gas-salt interface, the amount of chlorine should reflect the levels expected in the gas atmosphere.

For a given temperature, pressure and relative amounts or activities of components it is possible to calculate the stable assemblage of phases and their compositions by minimising the Gibbs energy. If necessary the amounts of minor species are then calculated by equalisation of chemical potentials followed by re-evaluation of the mass balance. Table 5.2 shows the results of a calculation for the multicomponent system discussed below. The substances are grouped by phase. A number of less important gas phase species and condensed stoichiometric substances have been omitted from the table. The values of chemical potential printed towards the end of the table are expressed in accordance with SGTE practice [85Bar]. It is possible to use the chemical potential to test whether the calculated equilibrium mixture would react with materials not considered in the calculation e.g. for possible corrosion of the container. In accordance with the requirements of equilibrium the chemical potential of a species such as NaCl, that is present in more than one phase, is the same in all phases. The calculation shows that Cr_2O_3 does indeed dissolve in the salt phase potentially exposing the turbine blade to corrosive attack from the oxidising atmosphere.

Even under modest pressures of SO_3 most of the sodium chloride can be expected to be converted to sodium sulphate. At the temperature of 1023.15 K sodium sulphate itself is solid, whereas experimental evidence suggests that the corrosive agent at this temperature is a sodium sulphate rich liquid phase. However, Figure 5.2a shows that a liquid can form containing about 28 mol% NaCl, a little NaOH and the remainder Na_2SO_4 but this would depend on there being a sufficiently high partial pressure of HCl. The pressure of HCl required to cause the liquid to form depends on the partial pressure of SO_3 which in turn depends on the partial pressure of oxygen. Figure 5.3 shows the coexistence line between the liquid and crystalline sodium sulphate and also the less important line for the coexistence of liquid and the NaCl-NaOH of the halite structure.

The equilibrium between the liquid and crystalline sodium sulphate is governed by the chemical reaction

$$Na_2SO_4 + 2\ HCl = 2\ NaCl + SO_3 + H_2O \qquad (5.1)$$

The proportion of NaOH is greatest under conditions of low pressure of HCl and SO_3 as can easily be anticipated. The straightness of the line in Figure 5.3 is due to the fact that the concentration of NaOH is small and the proportions of NaCl and Na_2SO_4 in the liquid are insensitive to the relative pressures of HCl and SO_3, provided the liquid and crystalline sulphate coexist. However, as indicated by reaction (5.1), the position of the line and hence the conditions for liquid phase formation are dependent on the partial pressure of water. Thus the rates of diffusion of water in the crystalline and liquid phases may be important parameters in hot corrosion since these rates will influence liquid formation in microenvironments.

The interaction of gas and salt with Cr_2O_3
For systems of more than three components it is not possible to represent the results of calculations on a conventional phase diagram. Table 5.2 shows the results of an individual calculation in the eight component system

C–Cl–Cr–H–N–Na–O–S

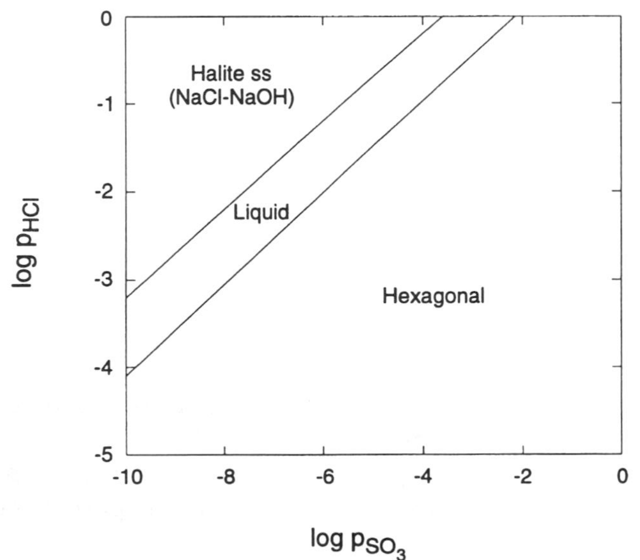

Fig. 5.3 Diagram showing the conditions for coexistence of the three component liquid phase NaCl-NaOH-Na_2SO_4 with the two-component halite phase and hexagonal Na_2SO_4 at 1023.15 K, p(H_2O) = 0.55 bar.

Table 5.2 Edited output from a single calculation of equilibrium in the system C-Cl-Cr-H-N-Na-O-S.

Temperature = 1023.1500 K
Fixed gas volume = 8.395500E-02 m3
Calculated gas pressure = 1.515009E+06 Pa

	Phase	Spec -ies	Amount mole	Mole fraction Partial pressure Notional activity
Phase GAS				
CClO<g>	2	18	6.463891E-18	6.464081E-18
CCl2O<g>	2	19	5.120377E-17	5.120528E-17
CHNO<g>	2	37	1.836733E-22	1.836787E-22
CO<g>	2	51	5.406494E-11	5.406654E-11
CO2<g>	2	52	6.137000E-01	6.137181E-01
Cl<g>	2	58	3.072932E-06	3.073023E-06
Cl2<g>	2	59	3.209431E-05	3.209526E-05
Cl2CrO2<g>	2	60	4.670889E-06	4.671027E-06
ClH<g>	2	61	6.422286E-03	6.422476E-03
ClHO<g>	2	62	1.645855E-06	1.645903E-06
ClNO<g>	2	63	2.024754E-07	2.024814E-07
ClNO2<g>	2	64	4.534128E-10	4.534262E-10
ClNa<g>	2	65	3.134771E-05	3.134864E-05
Cl2Na2<g>	2	66	3.689358E-06	3.689467E-06
ClO<g>	2	67	2.671773E-07	2.671852E-07
ClO2<g>	2	68	4.223573E-11	4.223697E-11
Cl2O<g>	2	69	1.636331E-12	1.636379E-12
Cl2OS<g>	2	70	3.322069E-19	3.322167E-19
Cl2O2S<g>	2	71	5.796471E-17	5.796641E-17
CrO2<g>	2	79	5.076084E-17	5.076233E-17
CrO3<g>	2	80	3.446742E-11	3.446843E-11
H<g>	2	81	3.292599E-14	3.292696E-14
H2<g>	2	82	6.348202E-11	6.348389E-11
H2N<g>	2	84	5.320189E-22	5.320345E-22
H3N<g>	2	86	8.794008E-19	8.794267E-19
HNO<g>	2	88	1.275204E-12	1.275241E-12
HNO2<g>	2	89	3.443450E-07	3.443551E-07
HNO3<g>	2	90	1.057176E-08	1.057207E-08
HNaO<g>	2	92	4.931999E-10	4.932144E-10
H2Na2O2<g>	2	93	1.246164E-16	1.246201E-16
HO<g>	2	94	7.833063E-07	7.833294E-07
HO2<g>	2	95	6.286854E-08	6.287039E-08
H2O<g>	2	96	5.500873E-01	5.501035E-01
H2O2<g>	2	97	2.981711E-09	2.981798E-09
H2O4S<g>	2	98	1.127982E-09	1.128015E-09
N<g>	2	102	3.661168E-21	3.661276E-21
N2<g>	2	103	1.159968E+01	1.160002E+01
N3<g>	2	104	2.147879E-23	2.147942E-23
NO<g>	2	105	5.570745E-04	5.570909E-04
NO2<g>	2	106	8.112675E-05	8.112914E-05
NO3<g>	2	107	3.127762E-11	3.127854E-11

```
 N2O<g>               2 108  1.450395E-07  1.450437E-07
 N2O3<g>              2 109  2.759359E-13  2.759441E-13
 N2O4<g>              2 110  6.611005E-15  6.611200E-14
 Na<g>                2 112  7.544120E-16  7.544342E-16
 NaO<g>               2 114  7.364937E-16  7.365154E-16
 Na2O4S<g>            2 115  6.088666E-10  6.088846E-10
 O<g>                 2 116  4.590916E-10  4.591051E-10
 O2<g                 2 117  2.180927E+00  2.180992E+00
 O3<g>                2 118  4.169665E-11  4.169788E-11
 OS<g>                2 119  8.564493E-19  8.564745E-19
 O2S<g>               2 121  2.919684E-07  2.919770E-07
 O3S<g>               2 122  6.014627E-07  6.014804E-07
 Phase total is              1.495153E+01  1.495198E+01


Stoichiometric phases
 C                    1 1                  1.721567E-21
 C2Cr3                3 1                  0.000000E+00
 C3Cr7                4 1                  0.000000E+00
 C6Cr23               5 1                  0.000000E+00
 CNa2O3               7 1                  1.189317E-06
 Cr                   8 1                  4.369075E-23
 CrN                  9 1                  1.261701E-20
 Cr2O3               11 1    3.210702E-02  1.000000E+00
 Cr2O12S3            12 1                  8.957969E-19
 CrH2Na4O6           25 1                  4.230024E-11
 HNa3O5S             27 1                  8.602759E-06
Phase LIQUID
 ClNa<LIQUID>        28 1    3.460268E-03  2.479055E-01
 CrNa2O4<LIQUID>     28 2    1.209187E-03  8.663033E-02
 HNaO<LIQUID>        28 3    2.027355E-07  1.452467E-05
 Na2O4S<LIQUID>      28 4    9.288352E-03  6.654496E-01
 Phase total is              1.395801E-02  1.000000E+00
Phase HALITE
 ClNa<HALITE>        29 1                  2.872939E-01
 HNaO<HALITE>        29 2                  5.161327E-07
Phase MONOCLINIC
 ClNa<MONOCLINIC>    30 1                  1.632411E-01
 HNaO<MONOCLINIC>    30 2                  4.444794E-06
Phase HEX
 CrNa2O4<HEX>        31 1    3.457211E-02  7.088038E-02
 Na2O4S<HEX>         31 2    4.531808E-01  9.291196E-01
  Phase total is             4.877529E-01  1.000000E+00
Phase ORTHO_NA2SO4
 CrNa2O4
  <ORTHO_NA2SO4>     32 1                  6.232106E-02
 Na2O4S
  <ORTHO_NA2SO4>     32 2                  2.551695E-01
Phase ORTHO_NAOH
 HNaO<ORTHO_NAOH>    33 1                  2.396193E-06
Phase NA2SO4
 Na2O4S<NA2SO4>      34 1                  2.458958E-01
```

Component	Ref. Phase	Chem. Pot.	Activity	Moles
C		-4.199367E+05	3.644258E-22	0.61370
Cl		-1.676515E+05	2.761495E-09	0.01000
Cr		-4.761963E+05	4.891850E-25	0.10000
H		-1.745120E+05	1.232839E-09	1.10660
N		-9.551675E+04	1.329623E-05	23.20000
Na		-3.567809E+05	6.106673E-19	1.00000
O		-1.098866E+05	2.455424E-06	8.22940
S		-4.833949E+05	2.098815E-25	0.46247
Total				34.72217

```
Helmholtz Energy = -4.3279448098E+06 J
```

Compnt	Phase	Mole fraction of compt within phase moles C	Cl	Cr	H
3.1067E+01	GAS	0.01975	0.00021	0.00000	0.03562
1.6054E-01	Cr2O3	0.00000	0.00000	0.40000	0.00000
8.0404E-02	LIQUID	0.00000	0.04304	0.01504	0.00000
3.4143E+00	HEX	0.00000	0.00000	0.01013	0.00000

Compnt	Phase	Mole fraction of compt within phase moles N	Na	O	S
3.1067E+01	GAS	0.74677	0.00000	0.19764	0.00000
1.6054E-01	Cr2O3	0.00000	0.00000	0.60000	0.00000
8.0404E-02	LIQUID	0.00000	0.30416	0.52224	0.11552
3.4143E+00	HEX	0.00000	0.28571	0.57143	0.13273

The pressures of N_2, CO_2, H_2O and initially O_2 were set to the values expected in the gas turbine environment. The amounts of sodium and chromium were fixed and the oxygen and sulphur proportions were adjusted to explore the range of conditions that might be found within the salt mixture at its contact with Cr_2O_3. The method allows the user of the system to calculate the partition of chromium between the four phases in which it is involved, namely Cr_2O_3, the crystalline solid solution of Na_2CrO_4 with Na_2SO_4, the liquid and the gas. The effect of the reducing conditions that might be expected within the molten salt deposit were explored by varying the oxygen potential.

Because the binary system Na_2CrO_4–Na_2SO_4 behaves almost ideally, both for the solid phase as well as for the liquid solution, an appreciation of the equilibrium between Cr_2O_3 and these phases can be obtained by considering the chemical reaction

$$Na_2SO_4 + 0.5\, Cr_2O_3 + 0.25\, O_2 = Na_2CrO_4 + SO_2 \tag{5.2}$$

The reaction suggests that the equilibrium between Cr_2O_3 and Na_2CrO_4 at a particular mole fraction in the crystalline phase can be plotted on a conventional phase stability diagram as shown in Figure 5.4. The slope of the lines in Figure 5.4 is 1/4 in agreement with reaction (5.2).
Moreover the lines are nearly equally separated in accordance with the fact

that the activity of Na_2SO_4 and the activity coefficient of Na_2CrO_4 in the solid solution are essentially constant. Because of this behaviour it is possible to combine the lines of Figure 5.4 into a single relationship by plotting the logarithms of the concentration of sodium chromate in the solid solution as a function of log $p(SO_2)$ - 0.25 log $p(O_2)$. The result is shown in Figure 5.5.

The same procedure can be followed for the solution of sodium chromate in the liquid phase. Despite the fact that the proportion of NaOH, though small, varies substantially and that of NaCl slightly with the imposed pressures of SO_2, O_2 and HCl, very little deviation from the two lines drawn in Figure 5.5 was calculated to occur. Liquid does not form under all conditions of oxygen and sulphur dioxide pressure. However, the conditions for liquid formation are not very much altered by the proportion of Na_2CrO_4 in solution and hence can be predicted by reference to Figure 5.3.

In the chosen model of the liquid phase the alkalinity, often expressed as oxide activity, is represented instead by the activity of NaOH. In principle the solution of Cr_2O_3 as Na_2CrO_4 might be represented by any of the following reactions

$$Na_2SO_4 + 0.5\ Cr_2O_3 + 0.25\ O_2 = Na_2CrO_4 + SO_2 \qquad (5.3)$$

$$2\ NaOH + 0.5\ Cr_2O_3 + 0.75\ O_2 = Na_2CrO_4 + H_2O \qquad (5.4)$$

$$Na_2O + 0.5\ Cr_2O_3 + 0.75\ O_2 = Na_2CrO_4 \qquad (5.5)$$

In conditions of chemical equilibrium these reactions will all give consistent results. On the other hand, in the non-equilibrium conditions obtaining at the interface between the salt and the oxide, material needs to be transported to and from the salt-gas interface. Reaction (5.3) would provide a good summary of the process if sulphur dioxide were able to diffuse rapidly through the salt or reaction (5.4) if diffusion of NaOH and H_2O were more rapid. The rates of diffusion would depend not only on the diffusion coefficients but also on the activity gradients. The concentration of Na_2O is unlikely to be sufficient to act as a vehicle for maintaining the alkalinity of the melt by diffusion from the salt-gas to the salt-oxide surface, as required by reaction (5.5).

In principle it is not necessary to express the problem in terms of specific stoichiometric reactions; indeed this carries dangers of misrepresentation. The reactions have been used here merely to indicate the need to include data for all the constituents that are active in diffusion processes.

Limitations of the data and calculated results

The calculated phase equilibria are of course totally dependent on the data. The authors are reasonably satisfied with the data for the pure substances

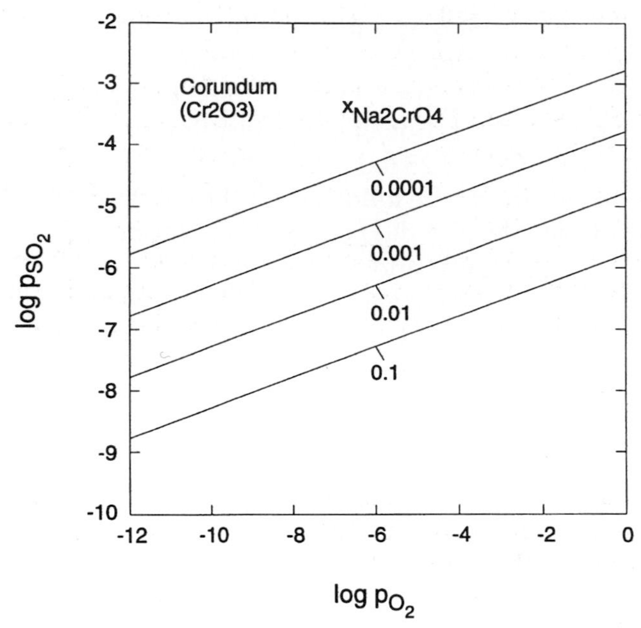

Fig. 5.4 The solubility at 1023.15 K of Cr_2O_3 expressed as the mole fraction of Na_2CrO_4 dissolved in the hexagonal Na_2SO_4 phase as a function of log p(O_2) and log p(SO_2).

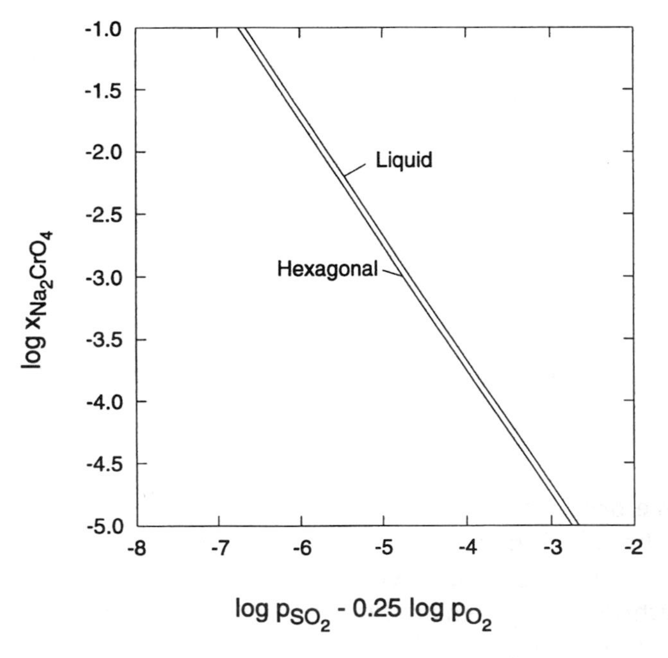

Fig. 5.5 The solubility at 1023.15 K of Cr_2O_3 expressed as the mole fraction Na_2CrO_4 in the liquid and hexagonal Na_2SO_4 phases. The conditions for existence of the liquid phase are indicated by Figure 5.3.

and solutions considered but it must be stressed that other substances or solution species need to be included. Thus under conditions of low oxygen potential sodium carbonate may become a significant component of the liquid. At high pressures of SO_3 sodium pyrosulphate ($Na_2S_2O_7$) may form and chromium may dissolve as the sulphate $Cr_2(SO_4)_3$. Furthermore it cannot be excluded that sodium chromite, $NaCrO_2$, may dissolve in the liquid phase. There is nothing in principle to stop the inclusion of these constituents except the limited amount of data available. The conditions for liquid phase formation would in practice be greatly extended by the presence of nickel and cobalt in the alloy [80Gup, 80Lut] and potassium in the gaseous and liquid environments. The calculations discussed above, therefore, are intended mainly as an indication of the power of recent thermodynamic methods.

Extension to higher order systems

The alloys commonly used in gas turbines contain the components nickel, cobalt and aluminium as well as chromium, and operate in an environment containing many others including nitrogen, oxygen, carbon, hydrogen, sulphur, chlorine, sodium, potassium, calcium, magnesium, phosphorus and vanadium. The oxides, sulphides and salts formed should be considered as solution phases as should the metallic and carbide phases which develop in subsurface regions of the alloy.

Coatings containing the additional elements, Si, Zr, Ti, and Y may also be applied and these will participate in all the phases already mentioned, including the alloy phases and the gas. It will be appreciated that the potential products of reactions are very numerous and a substantial volume of data is required. However, many of the same phases and systems that are important in gas turbine operation are also of interest in aqueous corrosion, pyro- and hydro-metallurgy, electronic materials, ceramics, catalysis and other applications of alloys. There is therefore the prospect that data can be accumulated from many sources, provided the assessments are based on consistent criteria.

In recognition of this fact there is a drive towards standardisation of the models and of key data values. Thus in many respects the methods and organisation required to provide the thermodynamic data needed for analysis of hot corrosion are already in place.

Future developments

To extend the calculations to the kinetics of chemical processes it is usual to assume that there is local achievement of equilibrium. With this assumption various authors have given analytical descriptions relevant to the oxidation of binary alloys with considerable success [83Laq, 83Nis]. However, for multicomponent alloys carrying multiphase corrosion products an analytical description is not feasible. Indeed, even individual calculations of chemical equilibrium use numerical and not analytical methods. The obvious method of modelling the kinetics of corrosion is therefore through the use of boundary element analysis. For this to be feasible data will be required in the form of coefficients of equations for the diffusion of species under the gradients of chemical (and perhaps electrical) potential present in the corrosion layers and in the depleted alloy. These coefficients will be temperature and composition dependent and will be required for each phase and for grain boundary diffusion.

If, for example, H_2O<g> is the agent that transfers oxygen along a crack to cause further oxidation, the calculation requires a knowledge of the interdiffusion coefficients of H_2O with all the other gases present. Provided all the data are available and provided the models and software for determination of chemical equilibria allow the chemical potentials and concentrations of diffusing species to be calculated, the user of the kinetic model does not need explicitly to state the processes involved in terms of chemical reactions. Thus it should not be necessary to decide in advance whether the solution of say nickel oxide in sodium sulphate occurs by alkaline or acid dissolution. The correct result will be automatically generated provided data for both models are incorporated into the database.

Further work will be required before a full thermodynamic analysis of the problem can be attempted. This could include the addition of some or all of magnesium, nickel, chromium, aluminium, potassium, vanadium, carbonate, chromite, pyrosulphate and oxide ions to the salt phase. Such ions could have the effect of lowering the melting point of the solid phase or providing a mechanism for the migration of chromium away from the surface of the turbine blade.

Acknowledgements

The authors gratefully acknowledge discussions with Dr. S. R. J. Saunders, concerning hot corrosion phenomena, and with Mr. R. H. Davies and Mrs. S. M. Hodson concerning various aspects of the data and calculation methods.

The calculations for this case study have been performed using MTDATA.

References

80Gup D.K.Gupta and R.A.Rapp: *J. Electrochem. Soc.*, 1980, 2194-2202.

80Lia1 W.W.Liang, P.L.Lin and A.D.Pelton: *High. Temp. Sci.* **12**, 1980, 41-50.

80Lia2 W.W.Liang, P.L.Lin and A.D.Pelton: *High. Temp. Sci.* **12**, 1980, 71-88.

80Lut K.L.Luthra and D.A.Shores: *J. Electrochem. Soc.*, 1980, 2202–2210.

82Bal C.W.Bale and A.Pelton: *Calphad* **6**, 1982, 255–278.

83Laq W.Laqua and H.Schmalzried: in 'High Temperature Corrosion', *NACE-6* (ed. R.A.Rapp) 115-120; 1983, Houston, National Association of Corrosion Engineers.

83Nis K.Nishida: in 'High Temperature Corrosion', *NACE-6* (ed. R.A.Rapp) 184–191; 1983, Houston, National Association of Corrosion Engineers.

85Bar T.I.Barry: in *Chemical Thermodynamics in Industry: Models and Computation,* (ed. Barry,T.I.) 1-39; 1985, Blackwell Scientific.

87Bar T.I.Barry and A.T.Dinsdale: *Mater. Sci. Technol.* **3**, 1987, 501–511.

6 Computer Assisted Development of High Speed Steels

PER GUSTAFSON*

Introduction

The optimisation of the composition of a high speed steel and the recommendation for heat-treatment are usually based upon a large amount of experimental work, because an adequate set of phase diagrams is not normally available for such a high order system. However, by assessing the lower-order systems thermodynamically it is possible to obtain a dataset from which the state of equilibrium can be calculated for any composition and temperature. Such a dataset has now been produced for the Fe–Cr–Mo–W–C system and it has been tested by comparing predictions with the results of some experiments. As a result, we can now trust the dataset and use it for compositions and temperatures different from the experimental ones. The experimental part of the development work can thus be reduced considerably.

Background

The earliest known high speed steels were tungsten base, 18:4:1 (W:Cr:V) being the best known. Different applications have resulted in the development of several variations on this composition, with W ranging from 22 wt% for increased red hardness and V to 5 wt% for greater resistance to abrasive wear. V dissolves in the MC_{1-X} carbide and will stabilise this otherwise rather unstable carbide. All these steels contain about 4 wt% Cr which is important in reducing scaling and in the hardening reaction, (increases the hardenability). More recently, molybdenum has entered this field. It can, due to the similarity between the two, replace W atom for atom and induce very similar properties.

Calculation

The following calculations will show a number of examples on how the description of the Fe-Cr-Mo-W-C system can be used to attain a better understanding of the properties of the high speed steel type of alloys. At the present, the databank contains a description of the Fe-Cr-Mo-W-C system. This description will in the future be extended to include other elements of

* R & D, Sintering Technology, AB Sandvik Coromant, S-12680 Stockholm, Sweden.

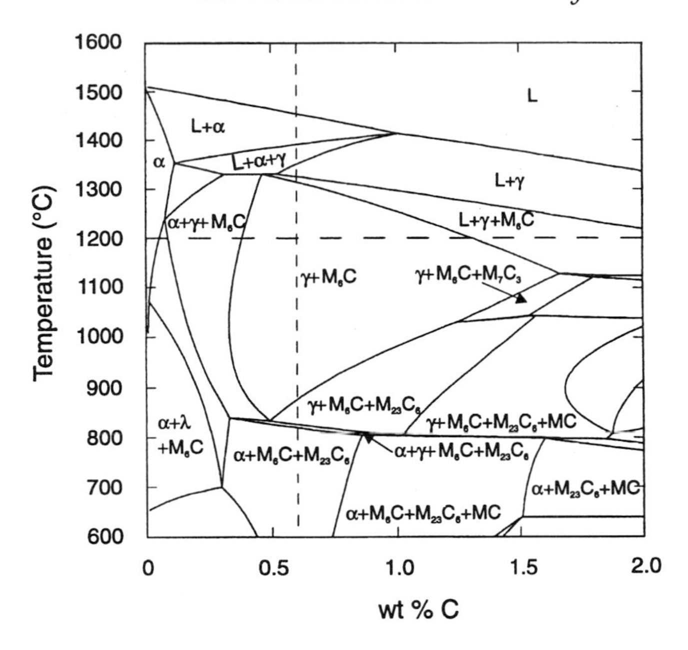

Fig. 6.1 Vertical section of the C-Cr-Fe-Mo-W system at 4 wt% Cr, 6 wt% Mo and 6 wt% W.

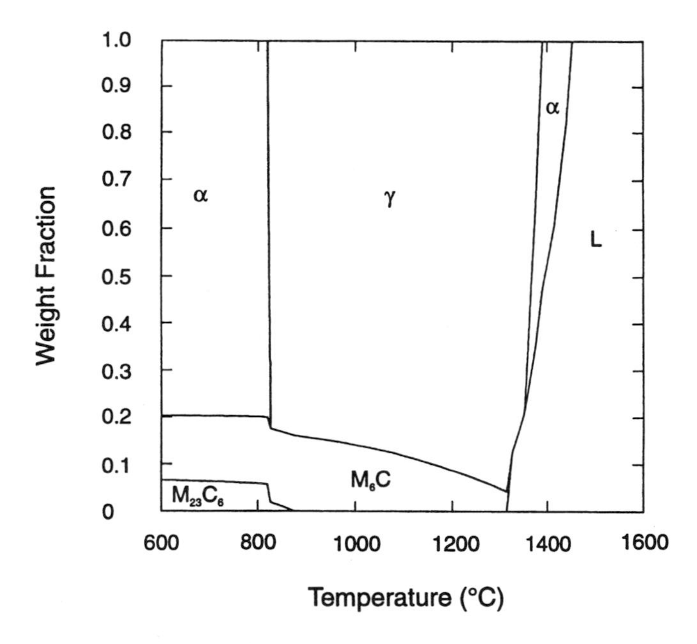

Fig. 6.2 The weight fraction of the phases forming in a 0.6 wt% C, 4 wt% Cr, 6 wt% Mo and 6 wt% W high speed steel as a function of temperature. The weight fraction of a phase is represented by the fraction of the vertical axis covered by the phase.

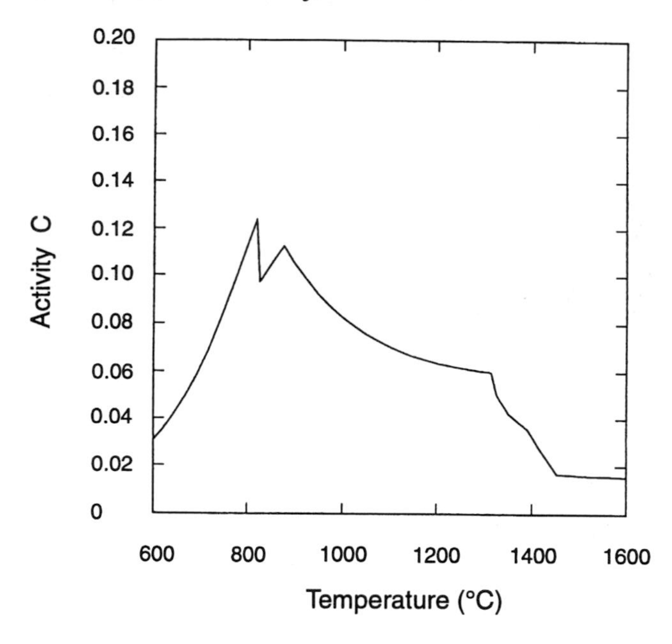

Fig. 6.3 The activity of C as a function of temperature for a steel of the composition given in Figure 6.2.

importance for the properties of many real high speed steels. The influence of Co and V was thus ignored in the calculations presented in this paper.

Discussion

Figure 6.1 shows a calculated vertical section in the C–Cr–Fe–Mo–W, system at 6 wt% W, 6 wt% Mo and 4 wt% Cr. In the past, various sections in the ternary C–Fe–W and quaternary C–Cr–Fe–W systems have been used as a basis for discussions on the properties of high speed steels. As an example the vertical dotted line shows the composition of a typical high speed steel, containing 0.6 wt% C. On cooling, this alloy will initially solidify to α, but that reaction will not go to completion because γ will start to form. The γ phase grows peritectically and all α will be consumed before the last liquid disappears and before the first M_6C carbide starts to form. At about 876°C some $M_{23}C_6$ will form and finally at about 820°C all γ will disappear due to the formation of α.

The calculated section alone does not give all information required. It may be more interesting to know how the amount of the various phases varies with temperature. This is better illustrated in Figure 6.2. This diagram shows the variation with temperature of the amount of the various phases that exist in a steel containing 0.6 wt% C. At each temperature the weight fraction of a phase is represented by the fraction of the vertical axis covered by the phase. Figure 6.3 shows how the activity of carbon changes with temperature in the same steel. It may also be of interest to know how a change in the composition influences the amount of the different phases at

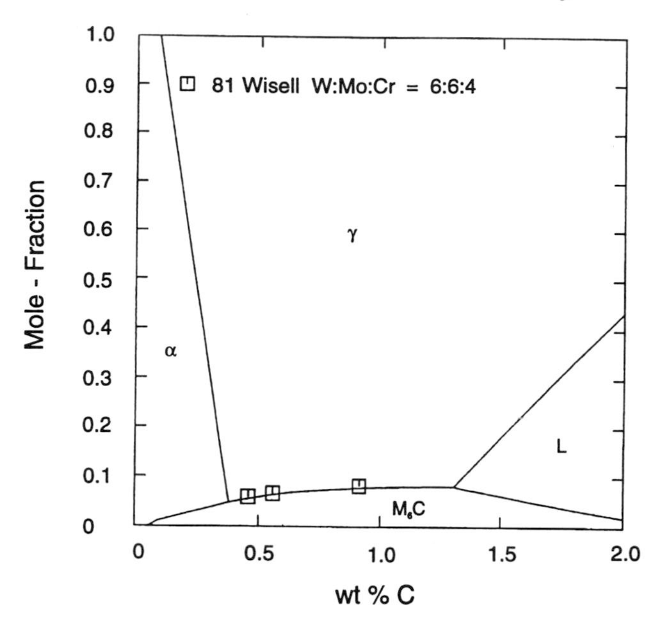

Fig. 6.4 The mole fraction of the phases forming in a 4 wt% Cr, 6 wt% Mo and 6 wt% W high speed steel at 1200°C as a function of the total carbon content. The weight fraction of a phase is represented by the mole of the vertical axis covered by the phase.

a given temperature. This is illustrated in Figure 6.4. This diagram shows how the amount of the various phases varies along the horizontal dotted line in Figure 6.1. The experimental data included in Figure 6.4 are taken from an unpublished work on high speed steels by Wisell [81Wis]. He also determined the carbon activity and the composition of the γ phase in the same alloys. Figure 6.5 shows the variation in the carbon activity along the same line in comparison with Wisell's data. Figure 6.6 shows the variation in composition of the γ phase along the line. The agreement between Wisell's experimental data and the present calculation is very good. This is encouraging especially since this information was not included in the evaluation of the thermodynamic properties of the C–Cr–Fe–Mo–W system.

The type of information given in the diagrams presented above can for example be of help to explain the effect of a given heat treatment or change in composition on the structure and mechanical properties of a high speed steel.

The calculations presented here are merely a few examples on what can be done with the data available in the SGTE databank.

The calculations for this case study have been performed using Thermo-Calc.

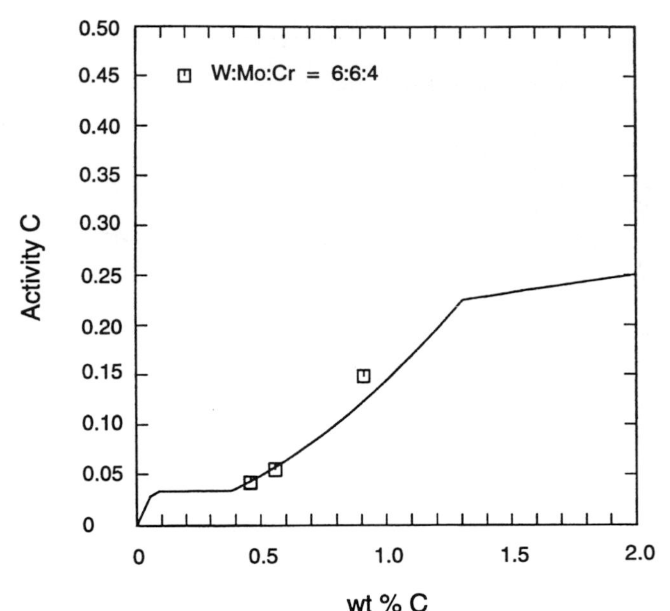

Fig. 6.5 The activity of C as a function of the total carbon content, for a steel given in Figure 6.4. The experimental data are from a previously unpublished work by Wisell [81Wis].

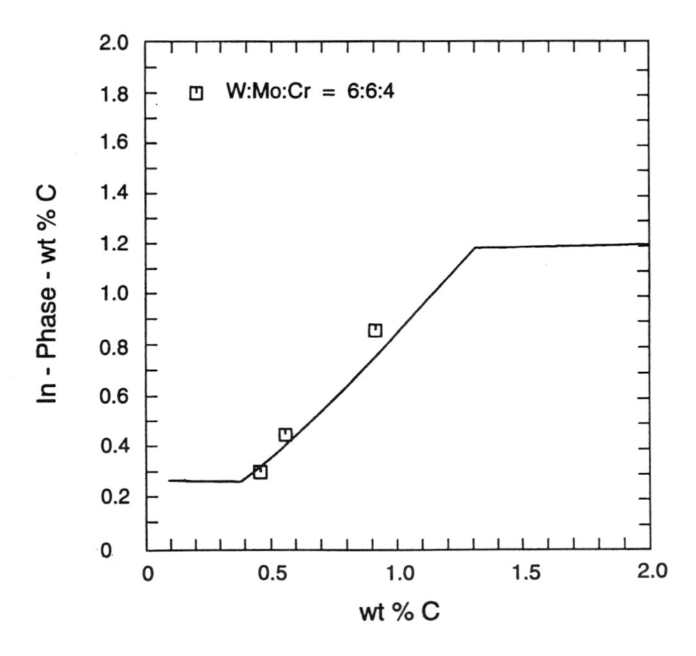

Fig. 6.6 (a) The composition of the γ phase as a function of the total carbon content for the steel given in Figure 6.4. (a) wt% C in γ as a function of the total carbon content.

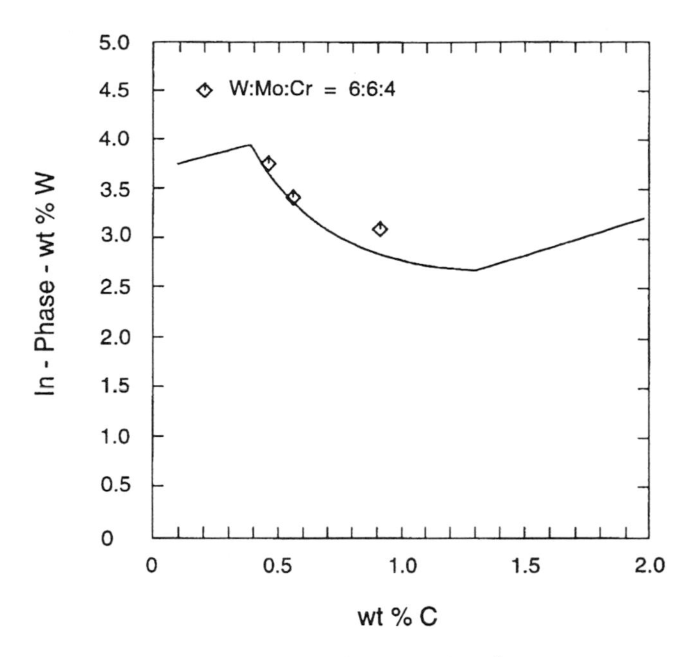

Fig. 6.6 (b) wt% W in γ as a function of the total carbon content.

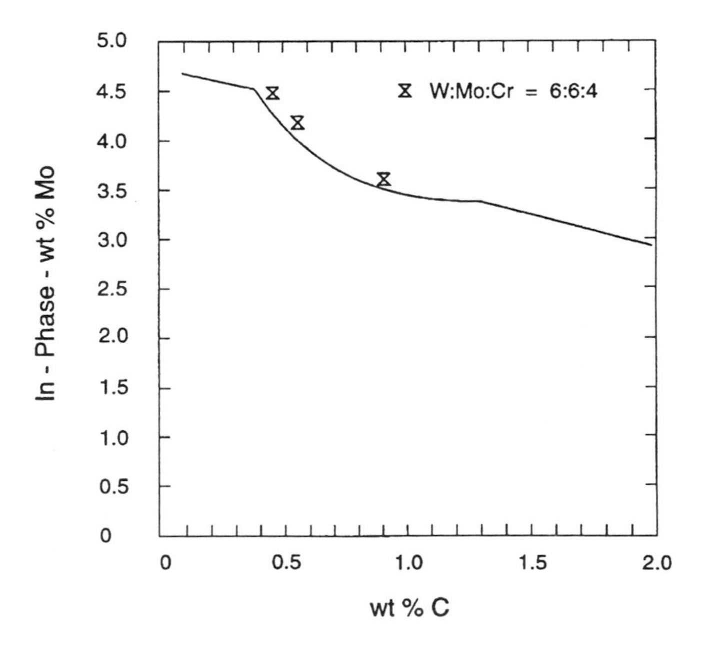

Fig. 6.6 (c) wt% M in γ as a function of the total carbon content.

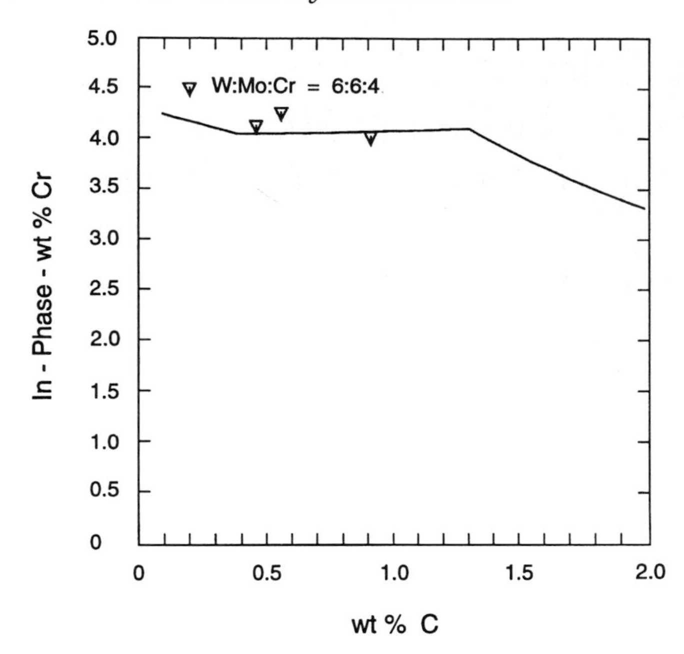

Fig. 6.6 (d) wt% Cr in γ as a function of the total carbon content.

References

81Wis H. WISELL: Laboratory report ASP-B 3/81, Kloster Speed Steel AB, Sweden.

7 Using Calculated Phase Diagrams in the Selection of the Composition of Cemented WC Tools with a Co–Fe–Ni Binder Phase

ARMANDO FERNÁNDEZ GUILLERMET*

Abstract

A thermodynamic description of the Co–Fe–Ni–W–C system is used to provide information of direct practical use in selecting the composition of (Co–Fe–Ni) bonded cemented WC tools. In particular, the problem of the optimum carbon contents is discussed. The limits of the range of carbon content in which only fcc and WC are formed on equilibrium solidification are identified for different compositions of the binder phase. In particular, the effects of replacing Co by Fe and Ni upon the identified region will be examined. The effect upon the solid liquid equilibrium will also be considered.

Background of the problem

It has long been known [54Gur, 70Mos] that the transverse rupture strength of Co-bonded WC tools is critically dependent upon the carbon content, and that the most favourable results are obtained when only fcc and WC are formed on cooling from the sintering temperature. For Co-bonded WC tools this is achieved if the W:C ratio is close to the value corresponding to the stoichiometric composition WC. An excess of carbon leads to the formation of graphite and a deficiency of carbon to the formation of the M_6C carbide. In this last case, the decrease in the rupture strength is particularly drastic.

When replacing Co by an Fe–Ni alloy the compositions leading to fcc + WC only do not necessarily coincide with the stoichiometric ratio. As a consequence early attempts to produce Fe–Ni bonded WC tools along the lines suggested by the experience with the Co based systems gave unsatisfactory results [60Sch], probably because of M_6C formation. More recent work [82Thu] on the effect of carbon upon the martensitic transformation in (Co–Fe–Ni) binder phases further highlights the need of information about the interesting range of carbon contents for WC tools.

It is shown here that this type of information can be obtained from calculated phase diagrams based on consistent thermodynamic datasets for all phases involved. The underlying assessment work has been described elsewhere [89Fer1].

* Consejo Nacional de Investigaciones Científicas y Téchnicas, Centro Atómico Bariloche, 8400 S.C. de Bariloche, Argentina

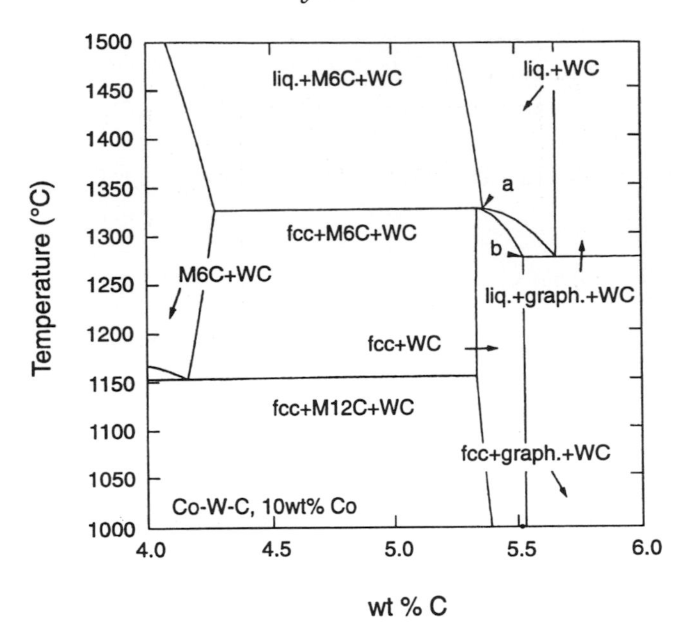

Fig. 7.1 Calculated vertical section (isopleth) of the Co–W–C system, constant Co content 10 wt% . Refer to text for explanation of solid symbol and points *a* and *b*.

The region of favourable carbon contents

Figure 7.1 shows a vertical section of the Co–W–C phase diagram calculated for a constant Co content of 10 wt%, a so-called isopleth diagram. It is evident from this graph that alloys with a carbon content falling between the compositions indicated by *a* and *b* will exhibit immediately after equilibrium solidification the two phases fcc and WC only. Of course consideration of the precipitation of the $M_{12}C$ carbide or pure graphite, which takes place at lower temperatures [89Fer2] will narrow the favourable range even further. However, neglecting these effects in the present discussion the range of carbon contents marked out by 'a' and 'b' shall be referred to as 'the favourable region'. A solid symbol on the weight percent axis of Figure 7.1 and all the following Figures indicate the composition of the respective system corresponding to contents of W and C in stoichiometric proportion to form WC. For the system shown in Figure 7.1 this composition falls within the favourable region.

Effects of replacing Co by Fe and Ni

Figures 7.2 and 7.3 show calculated vertical sections (isopleths) of the Ni–W–C and Fe–W–C phase diagrams. Comparison with Figure 7.1 elucidates the various effects of a full substitution of Co by Ni and by Fe respectively.

Ni moves the favourable region towards lower carbon contents compared with the stoichiometric composition without decreasing its width. Fe has

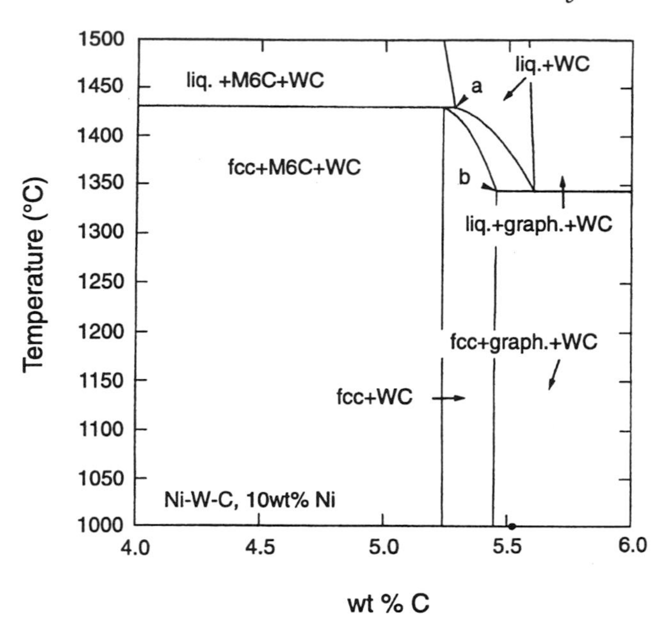

Fig. 7.2 Calculated isopleth of the Ni–W–C system, constant Ni content 10 wt%.

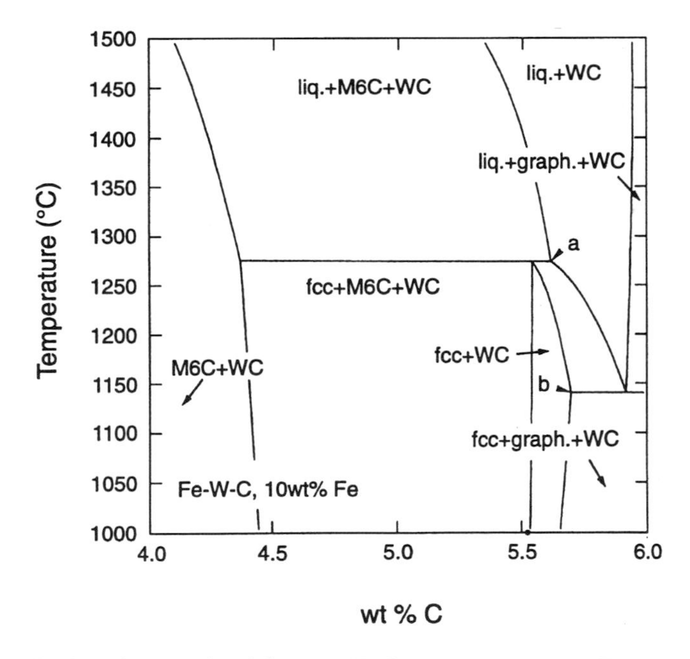

Fig. 7.3 Calculated isopleth of the Fe–W–C system, constant Fe content 10 wt%.

the opposite effect, moving the favourable region towards higher carbon contents and decreasing its width. Figure 7.2 shows also that the solid–liquid temperatures in the Ni–W–C system are higher than in the Co–W–C

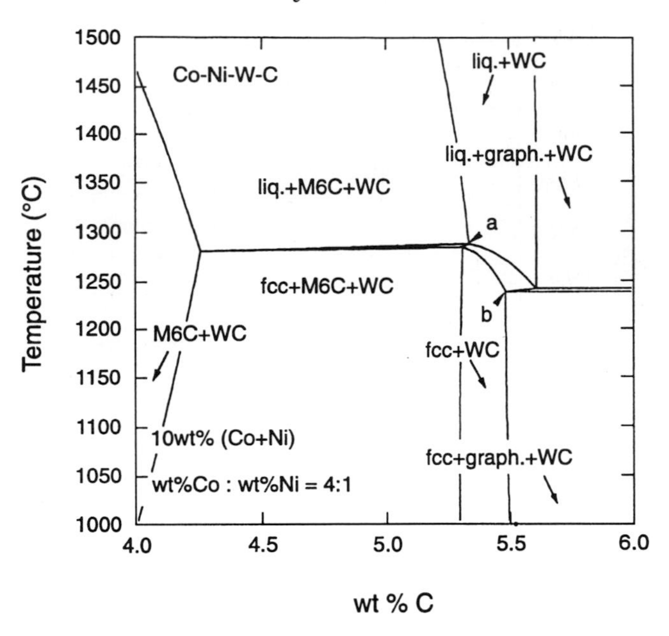

Fig. 7.4 Calculated isopleth of the Co–Ni–W–C system, constant (Co + Ni) content 10 wt%, wt%(Co) : wt%(Ni) = 4:1.

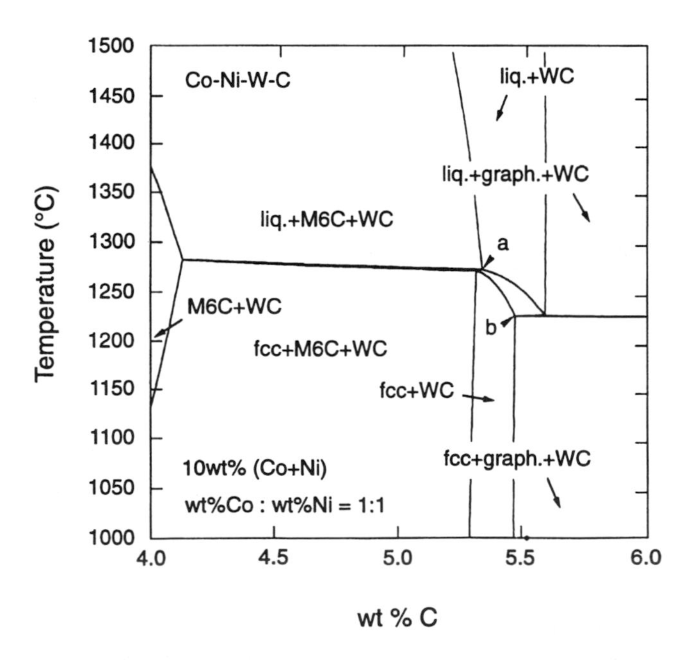

Fig. 7.5 Calculated isopleth of the Co–Ni–W–C system, constant (Co + Ni) content 10 wt%, wt%(Co) : wt%(Ni) = 1:1.

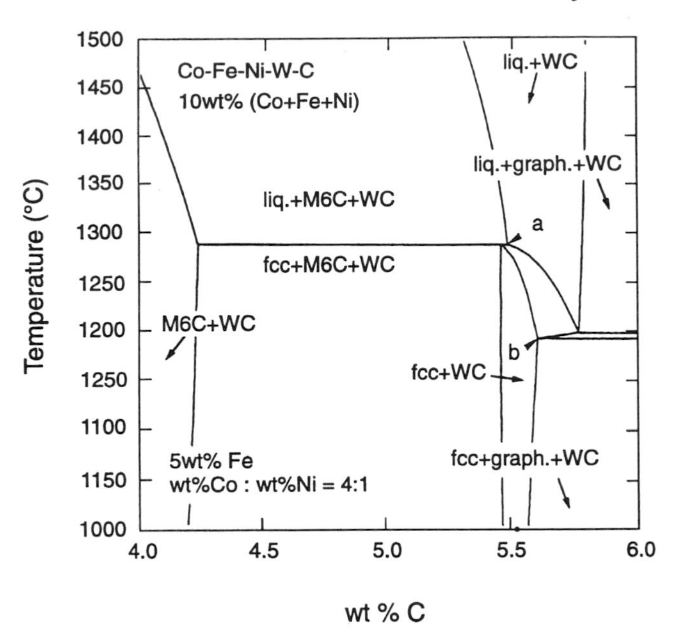

Fig. 7.6 Calculated isopleth of the Co–Fe–Ni–W–C system, constant (Co + Fe + Ni) content 10 wt%, constant Fe content 5 wt%, wt%(Co) : wt%(Ni) = 4:1.

system. However, the calculations predict that appreciable amounts of Co may be replaced by Ni without a drastic increase in the solid–liquid temperatures. This is demonstrated by Figures 7.4 and 7.5 which show sections (isopleths) of the Co–Ni–W–C system calculated for 10wt% (Co + Ni) and wt%(Co) : wt%(Ni) ratios of 4:1 (Figure 7.4) and 1:1 (Figure 7.5).

A question of practical interest is the replacement of part of the (Co + Ni) content of the system by Fe. Figures 7.6 and 7.7 illustrate the effect of replacing half the content of (Co + Ni) (in weight percent as given in Figures 7.4 and 7.5) by Fe. Compared with Figures 7.4 and 7.5 the calculated sections of the quinary Co–Fe–Ni–W–C system suggest relatively small effects of the replacement of (Co + Ni) by Fe. The favourable region is as expected somewhat narrower than in the Fe-free alloys with the same wt%(Co) : wt%(Ni) ratio and this unfavourable effect is more pronounced for the alloys with the lowest wt%(Co) : wt%(Ni) ratio. Besides, the substitution by Fe has moved the favourable region towards lower carbon contents. In Figures 7.6 and 7.7 this region is roughly centered around the stoichiometric composition.

Favourable carbon contents of a family of alloys

As an alternative to the use of single isopleths to determine the limits of the favourable region (i.e. the position of the points *a* and *b* in all previous graphs) of a set of specific alloys one can also study a family of alloys as was

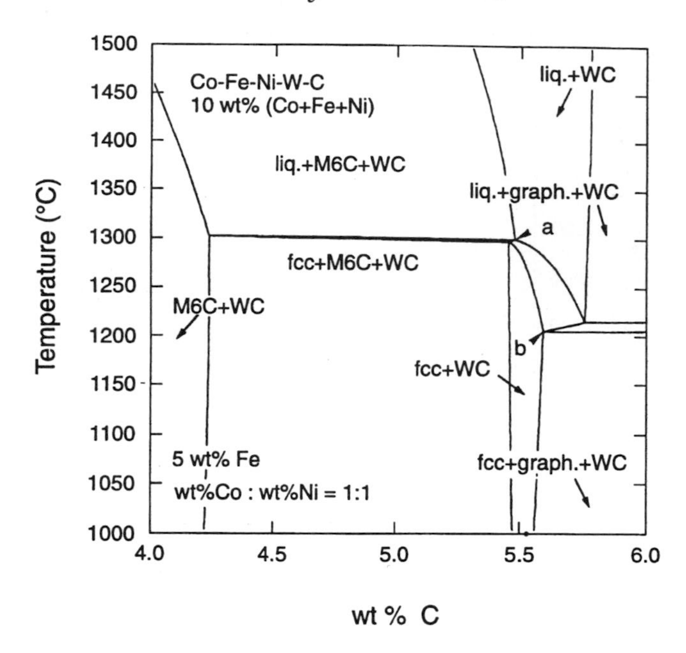

Fig. 7.7 Calculated isopleth of the Co–Fe–Ni–W–C system, constant (Co + Fe + Ni) content 10 wt%, constant Fe content 5 wt%, wt%(Co) : wt%(Ni) = 1:1.

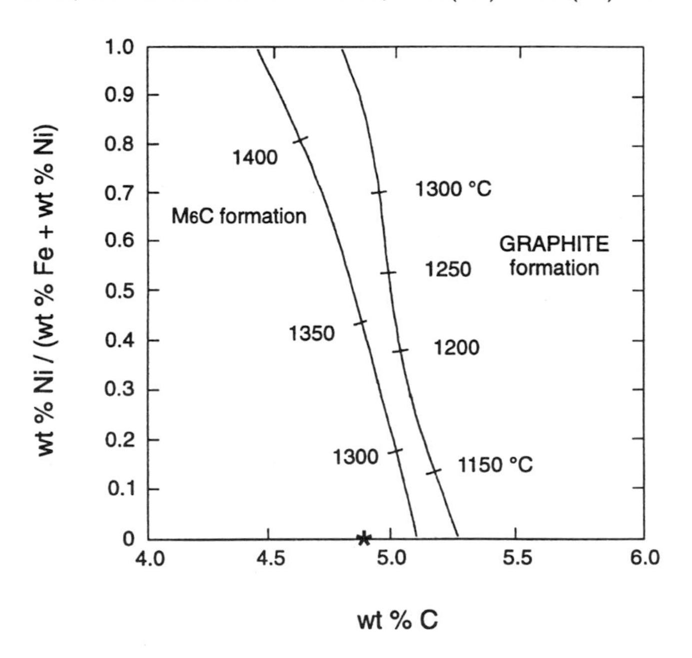

Fig. 7.8 Temperature projection of a section of the Fe–Ni–W–C system with 20 wt% (Fe + Ni). The lines describe the composition of a mixture of WC and liquid in equilibrium with fcc and M_6C (to the left), and WC and fcc in equilibrium with liquid and graphite (to the right). The asterisk on the composition axis represents the favourable region in the system.

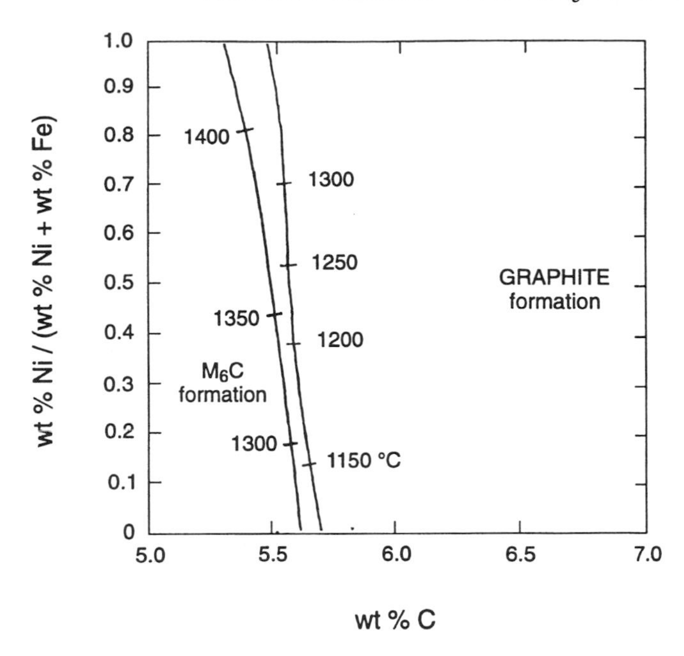

Fig. 7.9 Temperature projection of a section of the Fe–Ni–W–C system with 10 wt% (Fe + Ni). For explanation of the lines see Figure 7.8.

demonstrated for the Fe–Ni–W–C system in [87Fer1] and [87Fer2]. There it was shown that the lines defined by the displacement of the points *a* and *b* in the temperature composition space can be directly obtained by calculation. The results of such calculation for alloys of the Fe–Ni–W–C system with 20wt% (Fe + Ni) is demonstrated in Figure 7.8 which shows the temperature projections of the lines as a function of the wt%(Ni) : (wt%(Fe) + wt%(Ni)) ratio. The region between the lines in Figure 7.8 defines the favourable region of this family of alloys. If the (Fe + Ni) content is decreased the favourable region gets narrower. This is illustrated by Figure 7.9 which shows the result of the calculation for 10wt%(Fe + Ni).

Concluding remarks

The present case study demonstrates the possibilities of obtaining information of direct practical use in connection with (Co–Fe–Ni) bonded cemented WC tools by use of complex equilibrium calculations in multicomponent multiphase systems. By carrying out calculations based on the same set of thermodynamic data for other compositions ranges outside the ones covered by the present discussion can easily be covered.

The calculations for this case study have been performed using Thermo-Calc.

References

54Gur J.Gurland: *Trans. AIME* **200**, 1954, 285–290.

60Sch P.Schwartzkopf and R.Kieffer: *Cemented Carbides,* McMillan, New York 1960, 188–190.

70Mos D.Moskowitz, M.J.Ford and M.Humenik Jr.: *Int. J. of Powder Metallurgy* **6**, 1970, 55–64 (and references therein).

82Thu F.Thümmler, H.Holleck and L.Prakash: *High Temp.–High Press.* **14**, 1982, 129–141.

87Fer1 A.Fernandez Guillermet: *Zeitsch. Metallkunde* **78**, 1987, 165–171.

87Fer2 A.Fernandez Guillermet: *Int. J. of Refractory and Hard Metals* **6**, 1987, 24–27.

89Fer1 A.Fernandez Guillermet: *Zeitsch. Metallkunde* **80**, 1989, 83–94.

89Fer2 A.Fernandez Guillermet: *Met. Trans.* **A 20**, 1989, 935–956.

8 Prediction of Loss of Corrosion Resistance in Austenitic Stainless Steels

MATS HILLERT* AND CAIAN QIU*

Introduction

Austenitic stainless steels contain a combination of Cr and Ni which makes them fully austenitic under most practical conditions. The Cr content is usually about 18 wt% or higher which is well above the critical limit for corrosion resistance, about 12 wt% Cr.

It is difficult to produce commercial steels without any C and some decades ago it was usual to have around 0.1 wt% C in the austenitic stainless steels. This caused troubles in welded constructions because a Cr rich carbide may precipitate in the heat affected zone. The carbide forms at the grain boundaries by C diffusing there from the bulk material. Due to the slow diffusion of Cr, this element is only taken from a thin layer along the grain boundaries and that layer may thus be drastically depleted of Cr and may lose its corrosion resistance.

Clearly, the risk of losing the corrosion resistance would be lower if the C content were lower, and today it is possible to produce austenitic stainless steels with as low as 0.02 wt% C. However, for any C content there is a critical temperature below which a heat treatment can produce a depleted zone with less than 12 wt% Cr.

It would be of considerable practical value to be able to predict the critical temperature of any composition, i.e. for any combination of Cr, Ni and C contents. Such predictions will now be presented.

Theory

An accurate calculation of the formation of the depleted zone would require a detailed consideration of the diffusion-controlled growth of the C rich carbide. However, a rough estimate can be made already from the thermodynamic information on the Fe–Cr–Ni–C system, using the following method.

For chosen values of the temperature and Ni content, the C activity in the reaction zone along the grain boundaries can be calculated from the carbide/austenite equilibrium assuming that the C content of the austenite will be at the critical value of 12 wt%.

* Dept. of Materials Science and Engineering, Royal Institute of Technology, S–10044, Stockholm, Sweden.

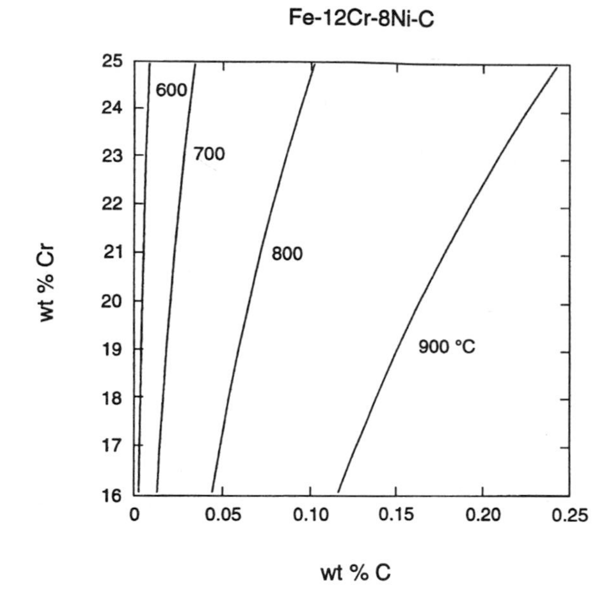

Fig. 8.1 Conditions for the formation of a depleted zone with a Cr content of 12 wt% in steels with 8 wt% Ni.

In view of the very large difference in diffusivity between C and Cr, it may be assumed that the rate of precipitation is controlled by the rate of Cr diffusion and C has sufficient time to diffuse over long distances and eliminate all differences in C activity. One may thus assume that the calculated C activity is also valid in the austenite far away from the grain boundaries and at an early stage of the reaction it will be equal to the initial C activity. One may then calculate the critical value of the initial C content of a steel, which will give a depletion of Cr down to 12 wt%.

In the above calculation it is necessary to make some assumption regarding the Ni content in the depleted zone of austenite. The Cr rich carbide will normally contain less Ni than the austenite and the Cr depleted zone will thus be enriched in Ni. A quantitative calculation of this enrichment requires a consideration of the rate of diffusion of Ni relative to Cr and is outside the scope of the present work. Hillert and Stawström [69Sta] chose to neglect this effect and simply assumed that the initial Ni content is also valid in the Cr depleted zone. This approximation will be accepted in the present work and the calculations of Hillert and Stawström will now be repeated but using an improved description of the thermodynamic properties of the Fe–Cr–Ni–C system.

Results

Figure 8.1 shows the critical C content in a steel with 8 wt% Ni as a function of the Cr content at four temperatures. For instance, in a steel with 18 wt%

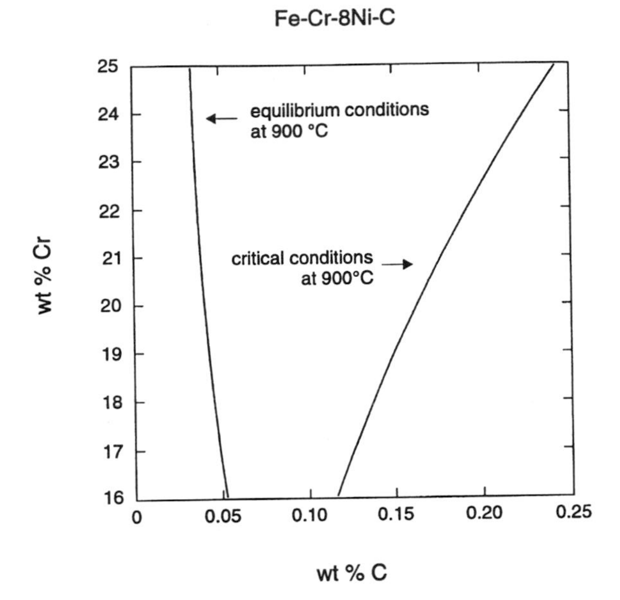

Fig. 8.2 Comparison between equilibrium conditions and conditions, where the Cr content in the depleted zone goes down to 12 wt%. The Ni content is 8 wt%.

Cr (and 8 wt% Ni) a carbon content of just over 0.05wt% will give the critical condition at 800°C. Below that temperature the Cr content in the depleted zone will be less than 12 wt%. These curves were calculated as described in the preceding section.

The slopes of the curves in Figure 8.1 show that more C can be tolerated if the Cr content is higher. This conclusion holds in spite of the fact that C and Cr both increase the tendency of forming the Cr rich carbide. This apparent contradiction is demonstrated for 900°C in Figure 8.2. The explanation is that Cr has a strong decreasing effect on the activity coefficient of C in austenite.

The relative positions in Figure 8.1 demonstrate a very strong effect of temperature. One must go to exceedingly low C contents in order to decrease the critical temperature below 600°C.

Figure 8.3 shows the critical C content in a steel with 18 wt% Cr as a function of Ni content at four temperatures. The slopes of these curves show that less C can be tolerated if the Ni content is higher. An increase from 8 to 24 wt% Ni decreases the tolerable C content to half.

Figures 8.4 – 8.6 show similar diagrams for higher Cr contents. The set of diagrams in Figures 8.3 – 8.6 covers the most common range of compositions in autenitic stainless steels.

Figure 8.7 shows the same kind of information plotted in a different way. Here the critical temperature is plotted as a function of the carbon content

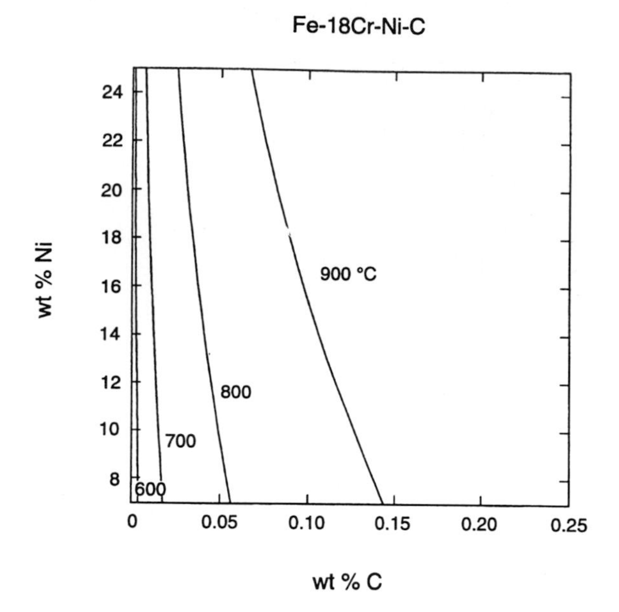

Fig. 8.3 Conditions for the formation of a depleted zone with a Cr content of 12 wt% in steels with 18 wt% Cr.

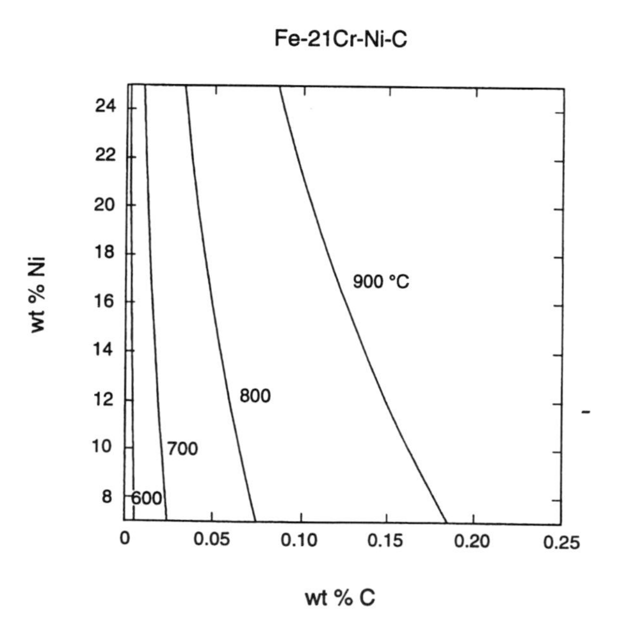

Fig. 8.4 Conditions for the formation of a depleted zone with a Cr content of 12 wt% in steels with 21wt% Cr.

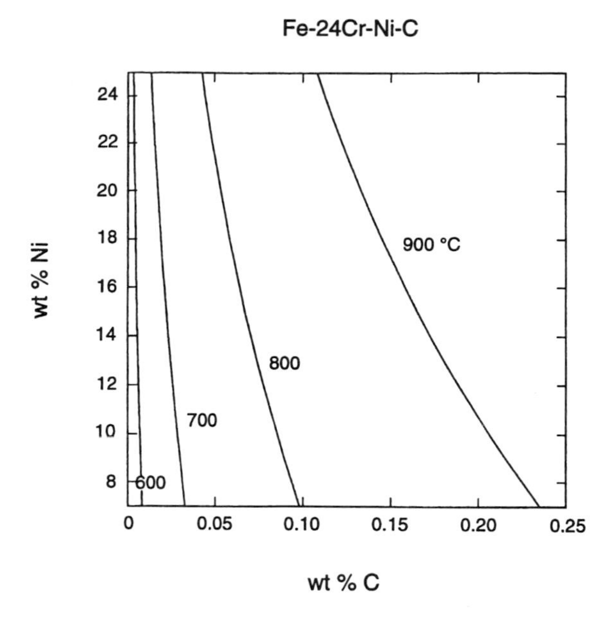

Fig. 8.5 Conditions for the formation of a depleted zone with a Cr content of 12 wt% in steels with 24 wt% Cr.

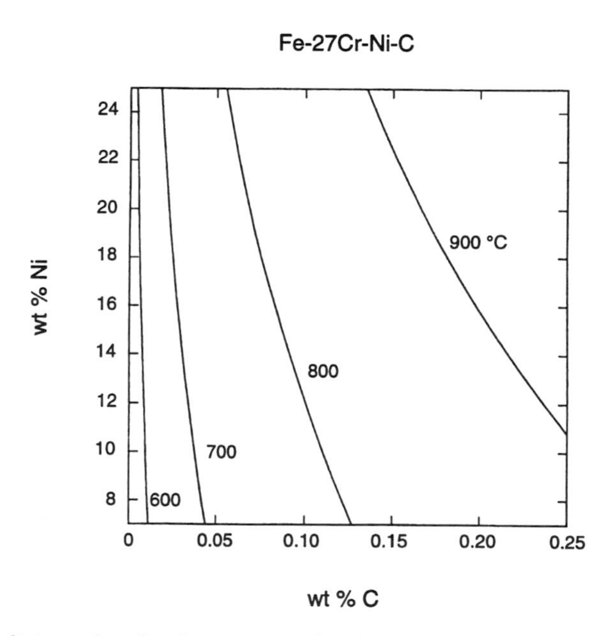

Fig. 8.6 Conditions for the formation of a depleted zone with a Cr content of 12 wt% in steels with 27 wt% Cr.

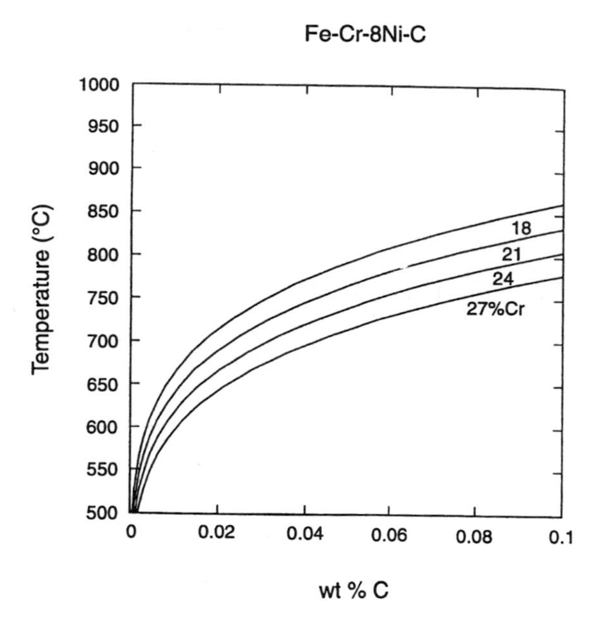

Fig. 8.7 The variation of the critical temperature, where the C content of the depleted zone goes down to 12 wt%, with the C content of steels with 8wt% Ni and 18 to 27 wt% Cr.

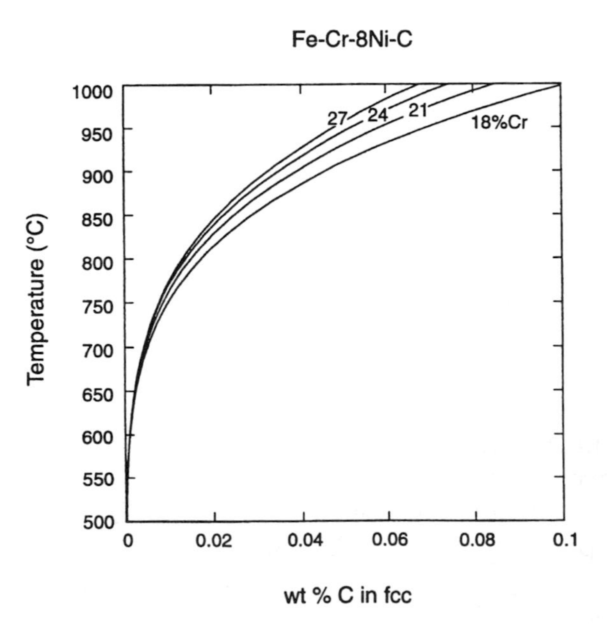

Fig. 8.8 The variation of the temperature, below which carbide starts to precipitate, with the C content of the steels considered in Figure 8.7.

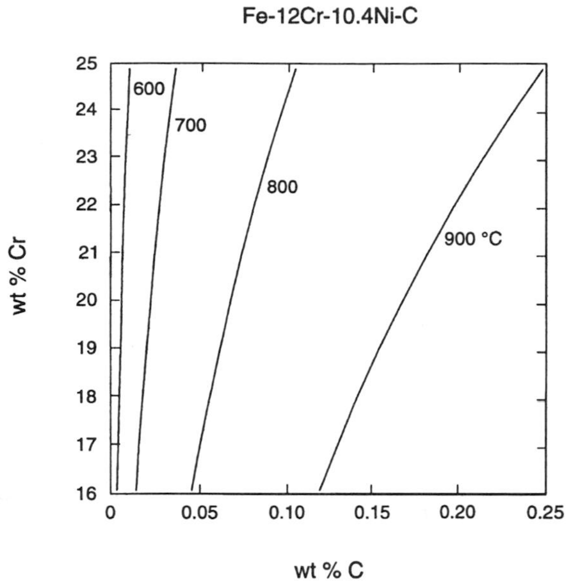

Fe-12Cr-10.4Ni-C

600
700
800
900 °C
wt % Cr
25 24 23 22 21 20 19 18 17 16
0 0.05 0.10 0.15 0.20 0.25
wt % C

Fig. 8.9 Conditions for the formation of a depleted zone with a C content of 12 wt% in steels with 10.4 wt% Ni.

for a number of Cr contents and 8 wt% Ni. It is again demonstrated that very low C contents are required in order to decrease the critical temperature to low values. For comparison, the equilibrium temperature is plotted in Figure 8.8. For 0.10 wt% C and 18 wt% Cr the difference is 130 K. For 27 wt% Cr it is twice as high because the higher Cr content increases the equilibrium temperature but decreases the critical temperature.

Of course, the curves in Figure 8.8 also show how high in temperature one must go in order to dissolve carbide which may have precipitated during a previous heat treatment.

Discussion

In the present work it has been assumed that the critical limit for the Cr content is 12 wt%. It should be realised that this critical limit most probably depends upon the composition of the steel. For instance, it seems to be well established that Mo may also contribute to the corrosion resistance and it seems most likely that Ni will also have an effect, maybe a negative one. As a consequence, the predictions presented in the present work should only be used as a guideline. The predicted effects of changing the Cr, Ni and C contents are probably more reliable and can be used if one wants to predict how a change in composition will affect the tendency of a steel to loose its corrosion resistance.

In an attempt to make the predictions more accurate, one may try to estimate how much higher the Ni content will be in the Cr depleted zone, compared to the bulk. A very rough consideration of the rates of diffusion of Cr and Ni showed that the local Ni content in the depleted zone may be higher by a factor of 1.3. In order to demonstrate the effect of such an enrichment, a series of calculations were carried out for a bulk content of 8 wt% Ni. See Figure 8.9 which should be compared with Figure 8.1. It is evident that this effect is almost negligible.

Method of plotting diagrams

The present calculations were carried out using the Thermo-Calc databank [85Sun]. It has a special feature for automatic calculation of diagrams involving two different but related equilibria in each step [90Hal] which can be illustrated by reference to Figure 8.3.

In this case each calculation is carried out in two steps. In the first step one defines the conditions of equilibrium between the carbide and austenite, by first giving the pressure (1 bar), temperature, Ni content (a value between 7 and 25 wt%) and Cr content (12 wt%) of the austenite and requiring that the carbide will be present but without giving its composition. The C content of the austenite required by the two-phase equilibrium is then calculated automatically and its C content is stored. In the second step a one-phase system of austenite is considered under the same pressure, temperature and Ni content, with the Cr content of the steel (18 wt%) and with the C activity obtained from the first step. Its C content is then calculated and is tabulated together with the Ni content used. The procedure is repeated with another Ni content until the range of Ni contents has been covered. Then the whole calculation is repeated for a new temperature. Finally, the content of the Table is plotted.

The other diagrams could be calculated similarly because it is possible to define the conditions for each equilibrium calculation in any way and to transfer any condition or result of the first calculation into the conditions for the second one.

Database

The present calculations were carried out with a database for the Fe–Cr–Ni–C system which has recently been completed [90Hil] and included in the Thermo-Calc databank. For the present calculations information on austenite and $M_{23}C_6$ carbide were used.

References

69Sta C.O.Stawström and M.Hillert: *J. Iron Steel Inst.* **207**, 1969, 77–85.

85Sun B. Sundman, B.Jansson and J.-O.Andersson: *Calphad* **9**, 1985, 153–90.

90Hal B.HALLSTEDT AND L.HÖGLUND: unpublished work, Royal Inst. Techn., Stockholm, 1990.

90Hil M.HILLERT and C.QUI: 'A thermodynamic assessment of the Fe–Cr–Ni–C system', *TRITA-MAC* **420**, 1990.

9 Calculation of Solidification Paths for Multicomponent Systems

Bo Sundman* and Ibrahim Ansara‡

Introduction

For solidification purposes in multicomponent systems, the crystallisation sequences which occur upon cooling are important for the properties of the material. The nature and the compositions of the various phases which precipitate can affect casting properties, microstructures, hence mechanical properties. It is difficult to determine these sequences by microanalysis because quantities such as partition coefficients are difficult to measure. In addition, the thermal effect associated with the process is a parameter which is difficult to evaluate experimentally.

Equilibrium compositions can provide such information if the thermodynamic properties of all the phases involved in the crystallisation process are known. Their compositions are generally derived by minimisation of the Gibbs energy of the system. Several algorithms exist for phase diagram calculations and will not be discussed here.

Gulliver–Scheil's model has been widely used in practice to simulate solidification for slow cooling rates. This model assumes local equilibrium at the solid–liquid interface, that the liquid is homogeneous and that there is no diffusion in the solid phases. Hence the fraction of the solid phase which precipitates will no longer participate in the solidification process.

During solidification, and as long as a two-phase equilibrium occurs, the overall composition of the system is continuously modified until a third phase precipitates. In this case, the composition of the liquid will follow a three or multi-phase line until no liquid remains.

The simulation of the solidification can be performed by considering stepping in temperature, or by fixing the proportion of the solid phase which precipitates or even by fixing the amount of energy which is extracted. Solidification of liquid Al–Mg–Si will be treated here and the examples which are given are relative to the case where temperature is decreasing.

* Department of Materials Science and Engineering, Division of Computational Thermodynamics, Royal Institute of Technology, S–10044 Stockholm, Sweden.
‡ Laboratoire de Thermodynamique et de Physico-Chimie Métallurgiques, URA CNRS 29, ENSEEG, BP 75, 38402 Saint-Martin d'Hères, France.

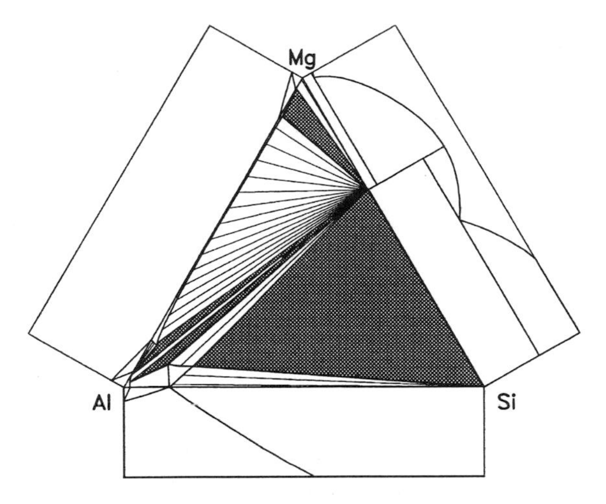

Fig. 9.1 Isothermal section of the Al–Mg–Si system at 580 °C.

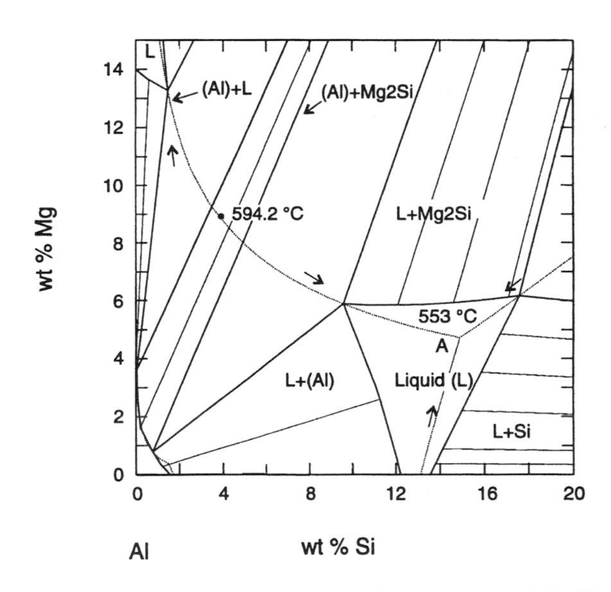

Fig. 9.2a Isothermal section of the Al–Mg–Si system at 853 K in the Al-rich region.

Description of the Phase Diagram

Figure 9.1 represents an isothermal section at 580 °C. There are no ternary intermetallic compounds in this system. The Al-rich corner is shown in Figure 9.2a on a larger scale. The dotted lines correspond to the monovariant lines. A maximum temperature equal to 594.2 °C occurs on the line representing the composition of the liquid phase in equilibrium with the Al solid solution and Mg_2Si. The eutectic temperature corresponding to the four-phase equilibrium between Si, (Al), Mg_2Si and the liquid phase is 553 °C.

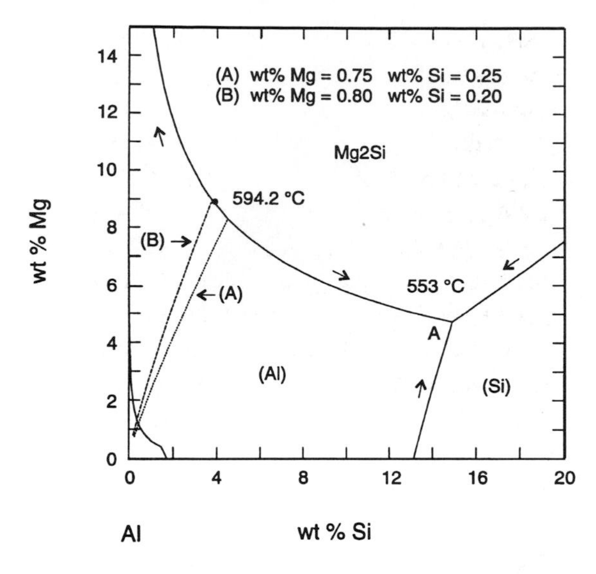

Fig. 9.2b Projection of the liquidus surface and solidification paths for two alloys.

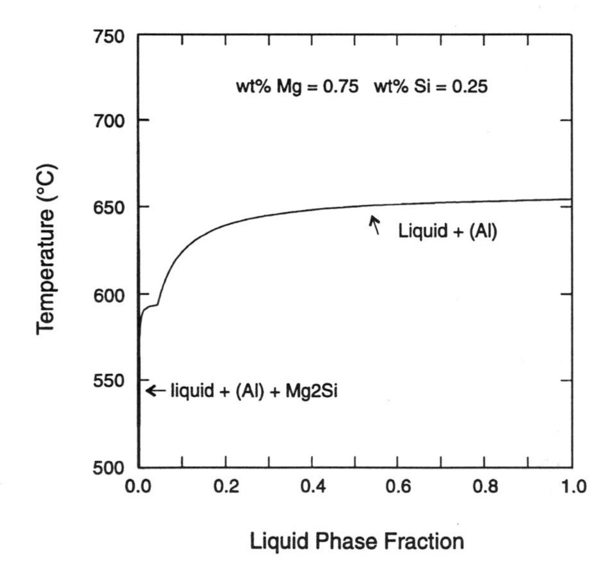

Fig. 9.3a Fraction of the liquid phase during solidification.

Solidification Paths

Figure 9.2b shows the solidification paths for two alloys for which the content of Mg and Si differs very little. The intersection of these lines with the monovariant line are on either sides of the maximum. Hence, the final alloy will present different microstructures, an eutectic structure containing silicon for the alloy (A) and Al_3Mg_2 for alloy (B).

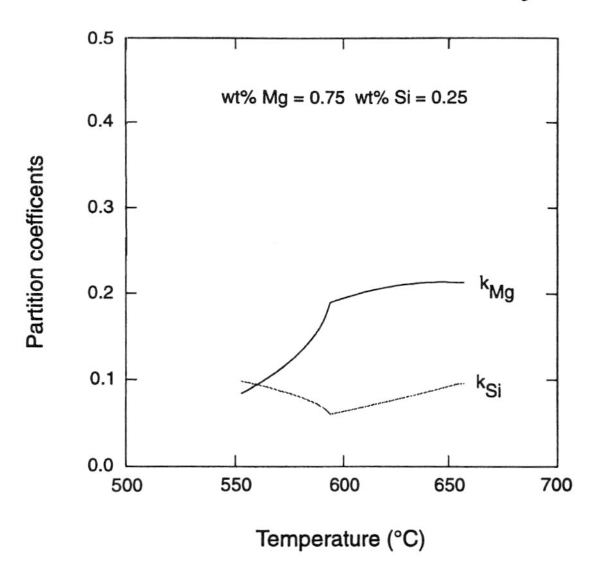

Fig. 9.3b Partition coefficient for Mg and Si during solidification.

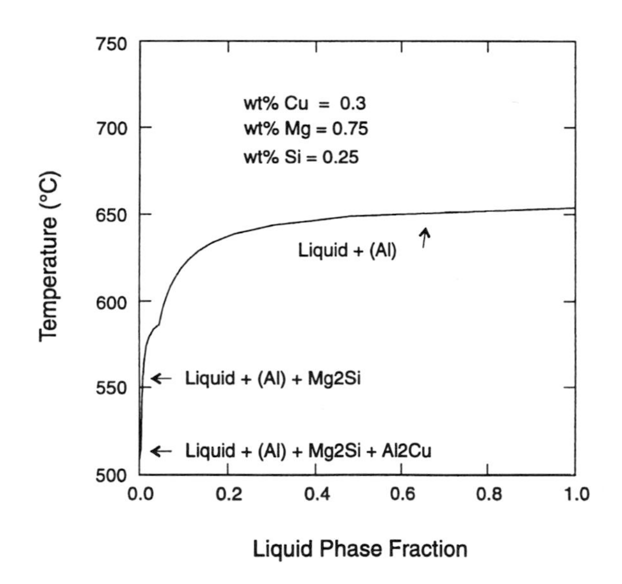

Fig. 9.4 Solidification path of an alloy containing 0.3 wt% Cu.

Figure 9.3a represents the fraction of the remaining liquid versus temperature for the alloy (B). The break point corresponds to the precipitation of the Mg_2Si.

In the Gulliver–Scheil's treatment, the partition coefficient is defined as the ratio of the composition of an element in the liquid and solid phases $k_i = x_i^l / x_i^s$, and is often assumed constant, even in multicomponent systems. This quantity is difficult to obtain experimentally because information on the phase composition is often lacking. As the phase diagram is calculated thermodynamically, a consequence is that the partition coefficients of all the elements in a multicomponent system can be derived. Figure 9.3b shows the evolution of the partition coefficients of Mg and Si versus temperature. If k_{Si} is constant, k_{Mg} varies slightly with temperature. Note that the partition coefficients change when the precipitation of Mg_2Si starts.

Similar characteristics are observed in a quaternary alloy where Cu is added, as shown in Figure 9.4.

The thermodynamic data used in these calculations are those existing in the SGTE (Scientific Group Thermodata Europe) database [86Ans]. The examples shown in this contribution have all been calculated using Thermo-Calc, a computer software for phase diagram and complex equilibrium calculations [85Sun].

References

85Sun B.Sundman, B.Jansson and J.O.Anderson: *Calphad* **9**, 2, 1985, 153–190.

86Ans I.Ansara and B.Sundman: 'Computer Handling and Dissemination of Data', Ed. P.S. Glaeser, Proc. Xth Codata Conf., Ottawa July 1986. Elsevier Sci. Pub. 1986.

10 Prediction of a Quasiternary Section of a Quaternary Phase Diagram

MATS HILLERT* AND STEFAN JONSSON*

Introduction

In order to visualise the phase diagram of a quaternary system under constant temperature and pressure one would need three dimensions. However, in some cases much information can be presented in a quasi-ternary section. This is the case for the Si–Al–O–N system where all the condensed phases fall in or very close to the section defined by the corners Si_3N_4, SiO_2, Al_2O_3 and AlN as if all the phases were ionic and composed of Si^{+4}, Al^{+3}, O^{-2} and N^{-3}, which is not quite true. It has been possible to model the properties of the condensed phases using models which restrict their existence to the section mentioned. The modelling will now be described briefly and it will be demonstrated that it was possible to extrapolate the properties of the ternary side systems and use the meagre information available from the quaternary system in order to predict the phase relations [91Hil2].

Solid Phases

There are two solid Si_3N_4 modifications, α and β, but their relative stabilities are not well known and it is difficult to separate information on their individual properties. In the present work the α phase was completely neglected. The β-Si_3N_4 phase extends far into the quaternary Si–Al–O–N system and, from a practical point of view, it is the most important phase in this system. The Si_2N_2O phase also extends into the system but has found less practical use. Inside the quaternary system these phases are often denoted by β′ and O′. In addition, there is a quaternary phase denoted by X. Its composition is not well known but was be treated as a stoichiometric phase with the composition $Si_{12}Al_{18}O_{39}N_8$. Finally, there is a whole series of so-called polytype phases related to AlN and situated close to the AlN corner. Their stabilities and properties are not well known and in the present work they were simply represented by one of them, the so-called 27R phase, which was treated as a stoichiometric phase on the AlN–Al_2O_3 side.

Modelling

It was assumed that the β′ phase can dissolve Al in the Si sublattice and O in the N sublattice. Accepting the normal valencies the formula was thus written

* Division of Physical Metallurgy, Royal Institute of Technology, S–10044 Stockholm, Sweden.

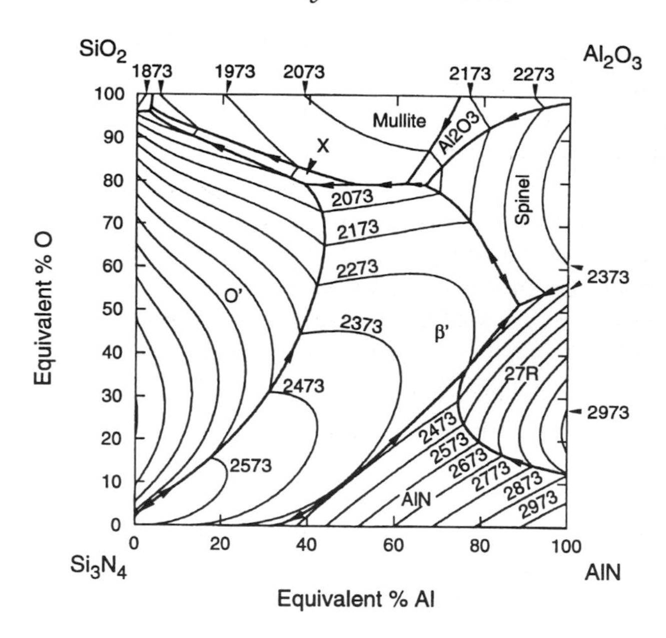

Fig. 10.1 Predicted projection of the liquidus surfaces with isotherms. The values give the liquidus temperature in kelvin.

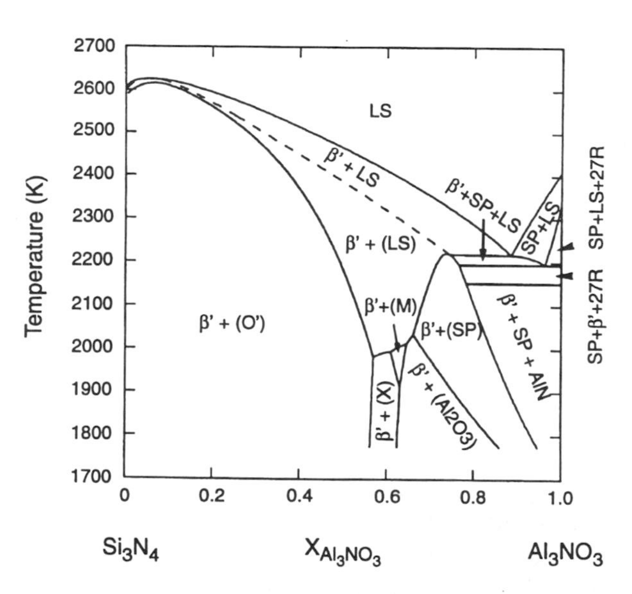

Fig. 10.2 Predicted section through the Si_3N_4–Al_3NO_3 line showing equilibria with β-sialon (β′). Two-phase fields with the other phase in parentheses indicate that the amount of the other phase is zero in the section. All regions below the dashed line actually represent 100 % β-sialon. The mullite and spinel phases are denoted M and SP, respectively.

as $(Si^{+4},Al^{+3})_3(N^{-3},O^{-2})_4$. A general type of model for phases with mixtures of elements on two different sublattices, called the compound energy model [70Hil, 86And], has been developed and programmed for computer calculations [85Sun]. In the present case a special procedure must be used in order to ascertain that the composition stays inside the quasiternary section. It is simply based upon the condition of electroneutrality. With this model the β′ phase extends along a straight line from the Si_3N_4 corner to the $AlN{\cdot}Al_2O_3$ point on the $AlN–Al_2O_3$ side.

The O′ phase was treated in the same way using the formula $(Si^{+4},Al^{+3})_2(N^{-3},O^{-2})_2(O^{-2})_1$. With this model the phase extends along a straight line from the Si_2N_2O point to the Al_2O_3 corner.

The modelling of the liquid phase is most difficult. In an attempt to describe the covalent character of the bonds and the tendency to form a network, the $SiO_2–Si_3N_4$ liquid was described simply by using hypothetical species SiO_2 and $SiN_{4/3}$ [91Hil1]. However, when analysing the properties of Al_2O_3 in various systems it was found that they could be represented with an ionic model, mainly using Al^{+3} and O^{-2} [90Hal]. The liquid phase in the quaternary system was finally described with the formula $(Al^{+3})_P(N^{-3},O^{-2},SiN_{4/3}{}^0,SiO_2{}^0,SiO_4{}^{-4})_Q$ using the ionic two-sublattice model [85Hil]. P and Q are coefficients to be adjusted to satisfy the condition of electroneutrality.

Results

The melting behaviour is particularly interesting in connection with the production and use of materials in this system. Figure 10.1 shows a projection of the calculated liquidus surfaces but it should not be trusted close to the AlN corner where a number of polytype phases may form. In the other regions of the system the diagram may be regarded as a reasonable prediction based upon extrapolation of the ternary properties from the four side systems and on the meagre information available inside the system.

A vertical section through the straight line of existence for the β′ phase is also of considerable interest. Figure 10.2 shows when β′ starts to melt (see dashed line) and also with what phases it may be in equilibrium in the case of a small excess of SiO_2 (in parenthesis). Phase fields, where β′ phase of a given composition decomposes into β′ of another composition and another phase upon heating, are given without parenthesis.

It is possible to combine the calculated properties of the solid phases inside the quaternary section with the known properties of the gas phase, which in general falls outside the section. In this way it was possible to produce a diagram (Figure 10.3) showing what pressure is required in order to prevent boiling before complete melting.

Even if the pressure is high enough to prevent boiling, there is a considerable risk of losing material by evaporation and thus change the

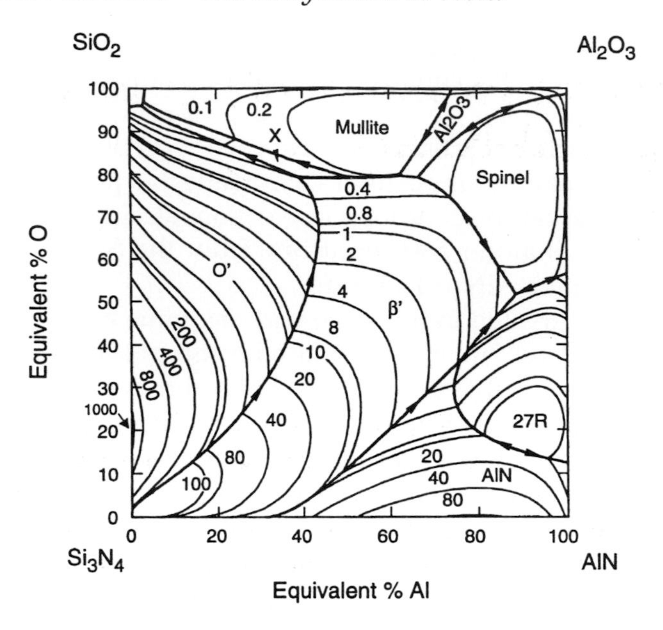

Fig. 10.3 Predicted projection of the liquidus surfaces with curves giving the vapour pressure in bar.

composition. The composition may thus move away from the quasiternary section which is in principle possible by the formation of metallic droplets.

The calculations for this case study have been performed using Thermo-Calc.

References

70Hil	M.Hillert and L.-I.Staffansson: *Acta Chem. Scand.* **24**, 1970, 3618–3626.
85Hil	M.Hillert, B.Jansson, B.Sundman and J.Ågren: *Metall. Trans.* **16A**, 1985, 261–266.
85Sun	B.Sundman, B.Jansson and J.-O.Andersson: *Calphad* **9**, 1985, 153–190.
86And	J.-O.Andersson, A.Fernández Guillermet, M.Hillert, B.Jansson, and B.Sundman: *Acta Metall.* **34**, 1986, 437–445.
90Hal	B.Hallstedt: *J. Am. Ceram. Soc.* **73**, 1990, 15–23.
91Hil1	M.Hillert and S.Jonsson: *TRITA-MAC* **465**, Royal Inst. Technology, Stockholm 1991.
91Hil2	M.Hillert and S.Jonsson: *TRITA-MAC* **470**, Royal Inst. Technology, Stockholm 1991.

11 Estimative Treatment of Hot Isostatic Pressing of Al–Ni Alloys

Klaus Hack[1]

Summary

Good data on the pressure related contribution of the Gibbs energy are still too scarce to calculate complete alloy phase diagrams for elevated pressures. However, it will be demonstrated that it is possible to obtain a first picture of the influence of pressure on a phase diagram by considering the effect of pressure on some univariant points, e.g. congruent melting points, eutectics or peritectics. For these equilibria it is possible to apply a generalised Clausius–Clapeyron equation, which only needs a few parameters.

Introduction

Although there are a number of processes with phase transitions taking place under elevated total pressure, such as in squeeze forging or hot isostatic pressing, metallurgists tend to neglect the influence of pressure on chemical equilibria since, in most cases, the pressure level is near or even below atmospheric pressure. While geologists have to treat the total pressure as one of their major variables and, therefore, have developed explicit expressions [44Mur] and consistent data bases [94Sax], calculations for alloy systems still have to be treated on an approximative level. Often, it is even difficult to find the values of the molar volume of an alloy phase, let alone the thermal expansion, the compressibility or the pressure derivative of the bulk modulus.

The following example demonstrates how such a 'thin' database can still be applied to provide a reasonable first estimate of the pressure dependence of alloy phase equilibria. However, it should be noted that the Clausius–Clapeyron equation presented in all standard text books for the case of a one component system first has to be generalised.

In the example, the pressure dependence of the peritectic temperature of the reaction $\gamma + \gamma' = liquid$ in nickel-based superalloys is calculated. The complex superalloy is approximated by the Ni–Al system, the phase diagram of which is given in Figure 11.1. The diagram shows that the phase

1 GTT-Technologies, D–52134 Herzogenrath, Germany

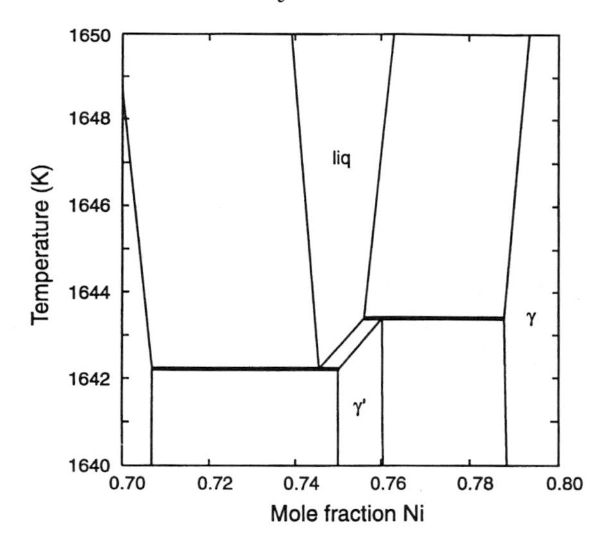

Fig. 11.1 Phase diagram of the Ni–Al system.

boundaries between the two solid phases run almost vertically. Since during HIP-ping the γ' phase is supposed to dissolve completely in the γ phase, the highest possible temperature is chosen. Theoretically this temperature is the peritectic. On the other hand the formation of liquid is to be avoided, because on solidification microporosity may occur, which is harmful to the mechanical properties. Thus the task is to find a value for the shift of the peritectic temperature on application of a pressure of 2000 bar at which the HIP process is operated.

Generalised Clausius–Clapeyron equation

Pelton and Schmalzried [73Pel] have shown that the phase boundaries in a phase diagram are mathematically governed by a Gibbs–Duhem equation with generalised potentials and conjugate extensive properties:

$$\sum Q_i^{\varphi} \mathrm{d}\phi = 0 \quad \text{with } \varphi = 1 \text{ to number of equilibrium phases} \tag{11.1}$$

With $\phi = T$, $-P$ or μ_i as potentials, S, V, and the molenumbers n_i must be used as the conjugate extensive properties. If the pressure dependence of an equilibrium has to be calculated, only the first two terms ($\Sigma S^{\varphi}\, \mathrm{d}T$ and $-\Sigma V^{\varphi}\, \mathrm{d}P$) apply as all changes in chemical potentials are zero at equilibrium. For a two phase equilibrium in a one component case the Clausius–Clapeyron equation is derived easily from :

$$S_1 \mathrm{d}T - V_1 \mathrm{d}P = S_2 \mathrm{d}T - V_2 \mathrm{d}P \; [=0] \rightarrow \frac{\mathrm{d}T}{\mathrm{d}P} = \frac{V_1 - V_2}{S_1 - S_2} \tag{11.2}$$

Note that this equation does not only apply to a phase transformation of an element, but to all two phase equilibria with one Gibbsean component, i.e. those cases in which both phases have the same composition with respect to **all** elementary components as stated by Gibbs [878Gib]. Examples are: Zn(s)=Zn(l), but also H2O(s)=H2O(l), CaO.Al2O3.2SiO2=liquid, or azeotropic points in organic systems. It is also worth mentioning that the absolute values of entropy and volume are used, i.e. the conjugate extensive properties to temperature and pressure respectively, although, because of the ratio, the result is the same if molar quantities are inserted.

For a three phase equilibrium such as a eutectic or peritectic in a binary system (binary in the sense of the phase rule), three equations of the above type are obtained. In the general case all three phases are solution phases and thus the composition in each phase may vary with pressure. Thus it becomes more obvious that the slope dT/dP is calculated for a certain fixed value $P°$. One obtains:

$$\frac{dT}{dP} = \frac{(x_3 - x_2)V_1 + (x_1 - x_3)V_2 + (x_2 - x_1)V_3}{(x_3 - x_2)S_1 + (x_1 - x_3)S_2 + (x_2 - x_1)S_3} \qquad (11.3)$$

with x_i the molefraction of component two in phase i and V_i, S_i the integral molar volume and entropy of the respective phase.

The numerator can easily be interpreted as the volume change $\Delta V(1 + 2 \to 3)$ (eutectic) or $\Delta V(1 \to 2 + 3)$ (peritectic) of the respective reaction with the appropriate interpretation of the entropy changes in the denominator.

A more general treatment of pressure dependence of all types of phase boundaries was given by Hillert [82Hil].

Application to the Ni–Al equilibrium

The data required for the application of the above equation to the Ni-rich equilibria in the Ni–Al system are summarised in Table 11.1. Thermochemical data are available from Murray [89Mur] who found eutectic behaviour between liquid, γ and γ', while Sundman and Ansara [90Sun] in a more recent assessment concluded a peritectic reaction. The majority of the volume data had to be based on assumptions (e.g. additivity of molar volumes of the component elements for the solution phases), interpolations or comparisons with similar metals. Different databases had to be used for the solid and the liquid state [75Tou, 90Lan]. It became obvious during the search for these data that there is a striking lack of precise data with respect to the pressure dependence for alloy calculations.

From the data in Table 11.1 the shift of the melting point of nickel (Ni(s)=Ni(l)) is calculated to be $dT/dP = \Delta V/\Delta S = 3.223 \times 10^{-8}$ K/Pa.

For the eutectic reaction according to Murray one obtains $dT/dP = 4.92 \times 10^{-8}$ K Pa^{-1}, while the peritectic reaction according to Sundman and Ansara leads to $dT/dP = 5.07 \times 10^{-8}$ K/Pa.

Table 11.1 Univariant phase equilibrium in the system Ni–Al.

(a) Pure nickel: Ni(s) = Ni(l), T_m = 1728.3 K

$\Delta V^{(1)}$ [m^3 mol^{-1}]	3.26×10^{-7}
$\Delta S^{(2)}$ [J mol^{-1} K^{-1}]	10.1139

(b) Eutectic reaction [89Mur]: $\gamma' + \gamma$ = liq, T_{eut} = 1658 K

	γ'	γ	liq
X_{Ni}	0.740	0.798	0.750
$V \times 10^6$ (1) [m^3 mol^{-1}]	8.126	7.891	8.71
S (4) [J mol^{-1} K^{-1}]	2.168	4.184	15.209

(c) Peritectic reaction [90Sun]: liq + γ = γ', T_{per} = 1643.4 K

	liq	γ	γ'
X_{Ni}	0.756	0.787	0.760
$V \times 10^6$ (1) [m^3 mol^{-1}]	8.658	7.928	8.037
S (4) [J mol^{-1} K^{-1}]	6.3774	−3.2833	−5.2587

(1) based on [75Tou, 90Lan]
(2) based on [91Din]
(3) relative to solid Al and Ni
(4) relative to liquid Al and FCC_A1 Ni

Thus, for a HIP pressure of 2000 bar = 0.2 GPa, the melting temperature of nickel is raised by 6.446 K, the three phase equilibrium by 9.84 K, or 10.14 K respectively. Consequently, under HIPping conditions, the maximum operating temperature may be chosen to be about 10 K higher than the annealing temperature at ambient pressure.

Conclusions

The generalised Clausius–Clapeyron equation yields a precise mathematical expression for the pressure derivative of univariant temperatures in binary (or higher order) systems. The type of the univariant equilibrium (peritectic or eutectic) and data for the entropy change of the reaction can be calculated for a good number of systems from different thermochemical assessments, while volume data are still very scarce. Nevertheless, for the Ni-rich part of the Ni–Al system the resulting pressure derivatives of the univariant temperatures can be used for reasonable estimates of temperature changes on application of technically feasible pressures.

Acknowledgements
The author wishes to thank Prof.D.Neuschütz, LTH, RWTH Aachen, for drawing attention to this subject matter.

References

878Gib J.W.Gibbs: 'On the equilibria of heterogeneous substances', *Proc. Acd. Connecticut*, 1878.
44Mur F.D.Murnaghan: *Proc. Natl. Acad. Sci.* **30**, 1944, 244.
73Pel A.Pelton and H.Schmalzried: Met.Trans. **4**, 1973, 1395.
75Tou Y.Touloukian et al.: *Thermophysical Properties of Matter*, Y.Touloukian (ed), vol. 11, New York, IFIPlenum, 1975.
82Hil M.Hillert: *Bulletin of Alloy Phase Diagrams*, **3(1)**, 1982, 4.
89Mur J.Murray: Private communication.
90Lan G.Lang: in *Handbook of Chem. Phys. Ref. Data*, 1990 Edition, CRC Press: Boca Ratton, FL.
90Sun B.Sundman and I.Ansara: COST507 database, 1990.
91Din A.T.Dinsdale: *Calphad* **15**, 1991, 317–425.
94Sax S.Saxena, N.Chatterji, Y.Fei and G.Shen: Thermodynamic Data on Oxides and Silicates, Springer Verlag, 1994.

12 The Thermodynamic Simulation in the Service of the CVD Process. Application to the Deposition of WSi_2 Thin Films

CONSTANTIN VAHLAS‡, CLAUDE BERNARD* AND ROLAND MADAR**

Introduction

Chemical vapour deposition (CVD) is one of the methods used for the production of thin films of different solid materials. It consists of depositing one or more solid phases on the surface of a heated substrate by means of a heterogeneous chemical reaction between the substrate and the feed gas phase, while the gaseous reaction by products are removed from the deposition area. Compared to other methods (sputtering, evaporation, etc.) CVD presents several advantages such as deposited microcrystalline structure, good step coverage, high throughput and possible selectivity with respect to substrate material.

The chemistry and the deposition mechanisms of the CVD are not always easy to control. Thus, although this method is in principle a very attractive one, it is rather difficult to optimise with respect to product yield, composition, surface morphology etc. A great number of preliminary experiments is often necessary in order to determine the optimum operating conditions, e.g. deposition temperature, total and partial pressures or nature and composition of the precursor gases. Many times the result still remains unsatisfactory.

An example illustrating this situation is the CVD of tungsten disilicide, WSi_2. WSi_2 is a potential candidate for the replacement of polycrystalline silicon (poly-Si) and aluminium as materials for integrated circuits in microelectronic technologies. Poly-Si and Al have long been used for gates and interconnecting conductor lines respectively in integrated circuits technology. Nevertheless, poly-Si has a relatively high resistivity which is greater than 500 μΩ cm even for a heavily doped material. This large electrical resistance can cause overheating and long response time delays in highly integrated circuits. On the other hand Al has a low electromigration barrier while its relatively low melting point considerably limits the permissible thermal treatments of the integrated circuit during the fabrication process. These problems were not very acute in the technologies applied so far. However, they substantially limit the new Very and Ultra Large Scale Integration (VLSI–ULSI) processes. An alternative to these limitations is to replace poly-Si and Al by a refractory metal silicide, such as

‡ Laboratoire de Cristallochimie, Reactivite et protection des Materiaux, CNRS-URA 445, ENSCT, 118 Route de Narbonne, 31077 Toulouse, France.
* LTCPM–ENSEEG, B.P. 75, 38402 Saint Martin d'Hères, France.
** LMGP-ENSPG, BP 46, 38402 Saint Martin d'Hères, France.

molybdenum-, tungsten-, titanium- or tantalum disilicides, ($MoSi_2$, WSi_2, $TiSi_2$, $TaSi_2$) or mixtures thereof. Moreover, the simultaneous replacement of two different materials by only one may reduce the number of steps necessary for the production of the circuit.

Many attempts have already been made for the CVD of WSi_2. From the reported results it appears that this is a very delicate task when the starting gas phase is composed of a mixture of tungsten hexafluoride, WF_6, silane, SiH_4 and hydrogen, H_2. The choice of this gas phase which is up to now the one most frequently used, is rather based on the physical properties or the availability of the constituent species than on the advantages they present from a thermodynamic point of view. For these reasons, WF_6, which is gaseous at ambient temperature was prefered to tungsten tetra- or hexachloride, WCl_4 or WCl_6, which are solids under the same conditions and are therefore more difficult to transport to the deposition area in a controlled and reproducible way. In the same way SiH_4 was initially prefered to other available silicon containing gases like dichlorosilane, SiH_2Cl_2 or disilane, Si_2H_6, since it is the commercially most common silicon precursor.

In order to understand the mechanism of the WSi_2 deposition and to improve the process, the study of some simple stoichiometric reactions has been suggested. These reactions assume in the deposition area the presence of only some of the many possible species at equilibrium. Consequently the results obtained may be inaccurate and some of the conclusions drawn may even be wrong. A thermodynamic investigation of the integral chemical system involved in a thermally (but neither plasma nor phonon) activated CVD process is far more appropriate to point out the influence of the experimental parameters on the characteristics of the deposited films in a systematic way. In the following, the different steps of this thermodynamic investigation will be presented through the simulation of the CVD of WSi_2 starting from fluorine or chlorine containing precursors.

Procedure

The first step in this direction is to inventory all the phases, stoichiometric or not, which are formed by combining the elements of the basic chemical system. The two chemical systems for the deposition from fluorides or chlorides are Si–W–F–H and Si–W–Cl–H respectively. Additionally, argon and oxygen have to be considered in the calculations since Ar will be used as a carrier gas and since it is interesting to monitor the presence of H_2O or traces of O_2 in the reactant gas mixture or the participation of the quartz made reactor walls in the deposition chemistry. The resulting chemical system is thus: Si–W–F(Cl)–H–O–Ar. In Table 12.1 are listed the generated gas species and condensed phases for the chlorine containing system.

The efficiency of the theoretical investigation is strongly dependent on the thermodynamic data applied. Bernard *et al.* [87Ber] illustrated the influence of the applied data on the quality of the resulting simulation by studying the deposition-etching regimes of Si on a Si substrate starting from

Table 12.1 List of phases and species for the Si–W–Cl–H–O–Ar system.

Species in the gas phase					
Ar	Cl	Cl_2	ClO	ClO_2	Cl_2O
H	H_2	HCl	HOCl	HO	HO_2
H_2O	H_2O_2	O	O_2	O_3	Si
Si_2	Si_3	SiCl	$SiCl_2$	$SiCl_3$	$SiCl_4$
SiH	SiH_2	SiH_3	SiH_4	Si_2H_6	SiH_3Cl
SiH_2Cl_2	$SiHCl_3$	SiO	SiO_2	W	WCl
WCl_2	WCl_3	WCl_4	WCl_5	WCl_6	WH_2O_4
WO_2Cl_2	$WOCl_4$	W_2OCl	WO	WO_2	WO_3
W_2O_6	W_3O_8	W_3O_9	W_4O_{12}		
Stoichiometric condensed phases					
Si	SiO_2(cryst)	SiO_2(quartz)	SiO_2(trid)	W	WCl_2
WCl_3	WCl_4	WCl_5	WCl_6	WH_2O_4	WO_2
$WO_{2.72}$	$WO_{2.9}$	$WO_{2.96}$	WO_3	WO_2Cl_2	$WOCl_4$
WSi_2	W_5Si_3				

a $SiCl_4$–H_2 mixture as a function of the Cl/H ratio in the reactive gas phase. They showed that attention must be payed to the interpretation of the obtained results since they may simply be due to a choice of inaccurate thermochemical data. Thus the first step towards the simulation of the process is to establish a correct and consistent set of thermodynamic data. If available, these data come from different internationally renowned thermochemical databanks such as the one of SGTE (Scientific Group Thermodata Europe). Otherwise the selection must be made directly from the information available in the literature.

The thermochemical description of the gas phase in the present systems is rather well known although the data of the species SiH_xCl_y seem to need a re-assessment in order to consider recent measurements by mass spectroscopy. On the other hand the W–Si binary system has not been assessed as a whole prior to the present calculations. Consequently the existing data of the different phases of this system and especially of the two compounds WSi_2 and W_5Si_3 which are stable under the common operating conditions of CVD showed great disparities. It was thus necessary to carry out a consistent assessment of the W–Si system. This was done by using standard optimisation techniques like the one of Lukas which takes

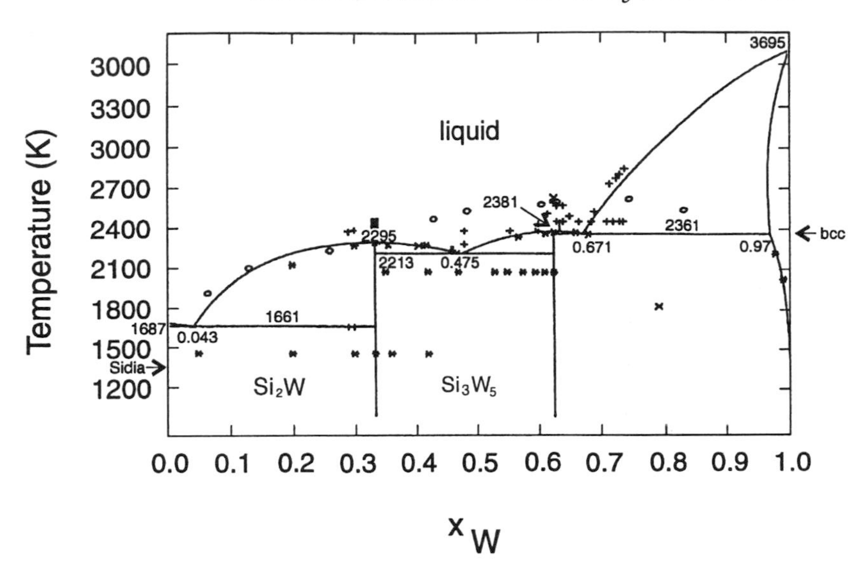

Fig. 12.1 W–Si assessed phase diagram.

- ○ R.KIEFFER, F.BENESOVSKY and E.GALLISTL: *Z. Metallk.* **43**, 1952, 284.
- × R.BLANCHARD and J.CUEILLERON: *Compt Rend. Acad. Sciences* **244**, 1957, 1782.
- + V.A.MAKSIMOV and F.I.SHAMRAI: *IZV. Akad. Nauk SSSR, Neorg. Mat.* **5(6)**, 1969, 1136.
- * YU.A.KOCHERZHINSKII, O.G.KULIC, E.A.SHISKIN and L.M.YUPKO: *Dokl. Akad. Nauk SSSR* **212(3)**, 1973, 643.
- # W.W.BEAVER et al.: Plansee Proc. 1964, *Metals for the Space Age*, Metallwerk Plansee AG Reutte/Tirol, 1965, 682.

intoaccount all the experimental information available, thermodynamic function values (enthalpies of formation, ΔH_f, heat capacities, C_p) as well as phase equilibria (fusion temperatures, Tf, three phase equilibrium temperatures and compositions)[89Vah]. The W–Si phase diagram which resulted from this work is shown in Figure 12.1. The corresponding thermochemical description of all the phases was integrated in the SGTE data bank. After this important preliminary step, the thermodynamic simulation of the CVD process could be undertaken. The thermodynamic investigation of the total chemical system is a complicated task and a computational method based on minimisation of its Gibbs energy is necessary to achieve it. For this purpose some software packages are available: for example Solgasmix and its new version ChemSage [76Eri] by G. Eriksson, the MultiPhase program [86Hod] by the National Physical Laboratory, Equi [75Ber] or Melange [86Bar] by the LTPCM. By these programs, the equilibrium values of the composition and amounts of the deposited phases and of the gaseous by-products are determined as functions of the reactive gas phase composition, the total pressure and the temperature.

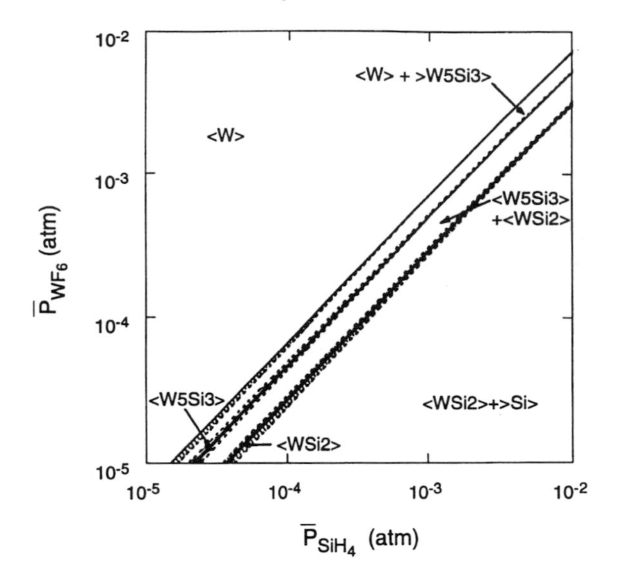

Fig. 12.2 CVD phase diagram of the WF_6–SiH_4–H_2–Ar system. $T = 1000$ K, $P_t = 1$ atm, $P_{Ar} = 0.9$ atm.

Different types of presentation can be used to illustrate the results of these thermodynamic calculations. Among them can be mentioned the so-called yield diagrams showing the number of moles or partial pressure or yield of one or more produced phases as a function of one of the operational parameters above. Another way of representation which was adopted in this study, is the CVD phase diagrams. They indicate the nature of the deposited phases (the tungsten silicides) as a function of two experimental parameters (x,y), most often the initial partial pressure of the reactive gases at a given total pressure and temperature. They are divided in different domains where same species are expected to be deposited.

As an example, in Figure 12.2 is presented the CVD phase diagram for the deposition from WF_6, SiH_4 and H_2 as a function of the SiH_4 and WF_6 initial partial pressure, for T = 1000 K, P_{tot} = 101.4 kPa and $P_{Ar} = 0.9.P_{tot}$. P_{H2} is obtained at each point as the difference between the total pressure and the sum of the above mentioned partial pressures. In this diagram two phase domains alternate with one phase domains which correspond to the W–Si phase diagram. The limits between two neighbour domains are shown rather as stripe type then as single lines. This comes from the fact that not only the values of the thermodynamic data themselves were taken into account during the calculations but also their uncertainties when available. The influence of these uncertainties on the results of the simulation is quantified by the width of the 'two dimensional' phase boundaries. The better the species involved are thermodynamically defined the narrower these boundary 'areas' are.

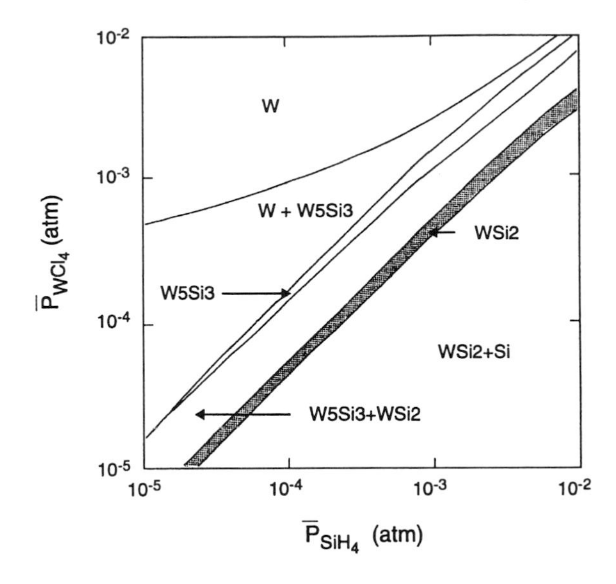

Fig. 12.3 CVD phase diagram of the WCl_4–SiH_4–H_2–Ar system. $T = 1000$ K, $P_t = 1$ atm, $P_{Ar} = 0.9$atm.

The results shown and discussed above come from the recent work of Thomas *et al.* [93Tho]. The authors adjusted the initial minimisation software in order to provide the width of the border 'areas' for every calculation. The systematic consideration of the precision of the used data as proposed by Thomas *et al.* increases the reliability of the thermodynamic simulation. It also indicates those species of the chemical system which need a more precise thermochemical description leading to the reduction of the width of the border 'areas' and consequently to the improvement of the simulation.

Results

From the study of Figure 12.2 it is easy to realise why the deposition of WSi_2 is such a delicate task under the conditions described: the deposition domain of pure WSi_2 is almost reduced to a line! The resulting CVD phase diagram when WF_6 is replaced by WCl_4 is presented in Figure 12.3. It is drawn for the same conditions as the previous one. The replacement of WF_6 by WCl_4 results in an increase of the width of the deposition domain of pure WSi_2. Consequently, a small fluctuation of the input gas composition will not necessarily result in the codeposition of the W_5Si_3 or Si phase. It thus seems a priori interesting to consider the use of chlorides instead of fluorides as tungsten precursors for the CVD of WSi_2. However, in that case the optimum conditions for transport of the tungsten chlorides have to be defined. This problem has been resolved [91Tho] by the use of the in situ chlorination of a thermoregulated tungsten bed positioned just above the deposition area.

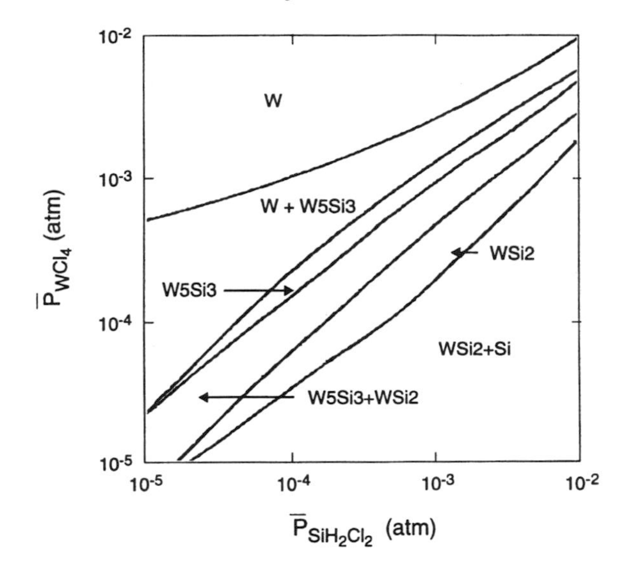

Fig. 12.4 CVD phase diagram of the WCl_4–$SiCl_2H_2$–H_2–Ar system. $T = 1000$ K, $P_t = 1$ atm, $P_{Ar} = 0.9$ atm.

Yet another variation of the process can also be checked by calculations with the same data set: If SiH_4 is replaced by SiH_2Cl_2 in the reactive gas phase the WSi_2 single phase deposition domain grows even further thus providing very comfortable conditions for the deposition of pure WSi2. This is illustrated in Figure 12.4 drawn for similar operating conditions as those for Figures 12.2 and 12.3.

Each one of the CVD phase diagrams presented concerns a specific gas mixture and the resulting conclusions cannot be extended to other systems even if they belong to similar chemical groups. Particular thermodynamic calculations are necessary for every specific initial composition in order to obtain the optimum deposition conditions. In Figure 12.5 the CVD phase diagram is presented [91Mas] for the deposition of titanium silicides from an SiH_4, H_2, Ar and titanium tetrachloride, $TiCl_4$ containing gas phase. It is drawn for the same operating conditions as for Figure 12.3. From the comparison of the two CVD phase diagrams it appears that the $TiSi_2$ single phase deposition domain is much bigger than the one for WSi_2 and consequently pure $TiSi_2$ is easier to obtain than WSi_2. High quality $TiSi_2$ thin films have already been obtained under these operating conditions in agreement with the diagram of Figure 12.5.

Limitations and further development

It was already mentioned that the thermodynamic simulation should in principle only be applied to the thermally activated CVD processes. Furthermore, to achieve a reliable simulation, the operation temperature

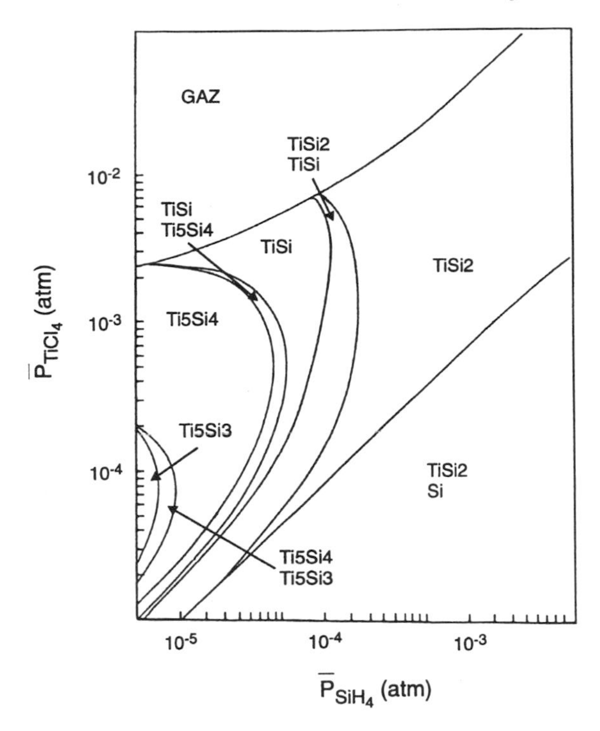

Fig. 12.5 CVD phase diagram of the $TiCl_4$–SiH_4–H_2–Ar system. $T = 1000$ K, $P_t = 1$ atm, $P_{Ar} = 0.9$ atm, $P_{H2} + P_{SiH4} + P_{TiCl4} = 0.1$ atm.

has to be high enough to be able to neglect surface kinetics in comparison to the gas phase diffusion. As a matter of fact, a diffusion limited process leads to results which are the closest to the thermodynamic predictions. This is shown in Figure 12.6 where Besmann and Spear [77Bes] present the growth rate for the CVD of titanium boride, TiB_2, as a function of temperature, starting from titanium tetrachloride, $TiCl_4$, boron trichloride, BCl_3, and H_2. Curves 1, 2 and 3 in this Figure correspond to deposition conditions controlled by kinetics, thermodynamic equilibrium and mass transport respectively. It appears that curve 2 forms the upper limit towards which tends the experimental curve 4. The form of curve 4 indicates that the evolution of the growth rate as a function of temperature may be divided in two domains: At low temperatures the growth rate depends strongly on T. Surface kinetics become predominant; the reaction rate is slower than the transport of the reactive gases into the deposition area. At high temperatures the growth rate is fairly stable. Deposition is controlled by gas phase diffusion; the reaction rate is too high compared to the transport of the reactive gases into the deposition area. Under extreme conditions (very high T and/or pressure or concentration) it may even lead to homogeneous nucleation [76Duc].

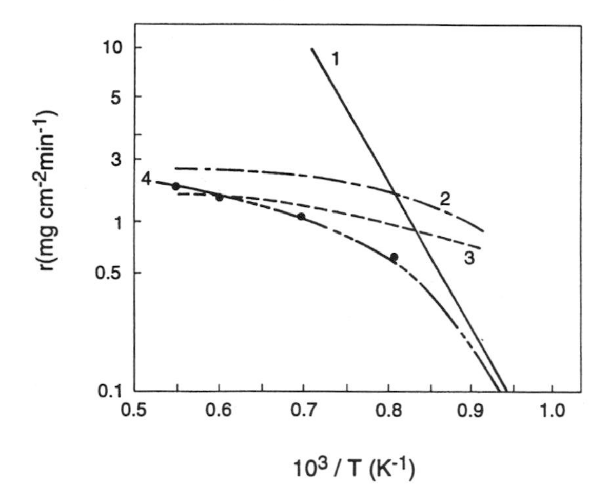

Fig. 12.6 Arrhenius plot for the deposition of TiB_2 from a mixture of $TiCl_4$, BCl_3 and H_2. This diagram includes observed rates (4) as well as calculated equilibrium-limited (2), diffusion-limited (3) and chemical kinetic-limited (1) rates.

From the above it appears that in many cases equilibrium calculations based on the gas phase composition immediately above the substrate surface can predict the composition of the deposited films as a function of the operating conditions. However, since in a CVD process not only thermal but also hydrodynamic phenomena occur, it is necessary, in order to predict and subsequently monitor the reactor performance (deposition rate, film uniformity and composition), to associate thermochemical and kinetic considerations. Additionally to the conservation of mass and the conservation equation for each species, the conservation of momentum and energy coupled with adequate boundary conditions must be taken into account. This complicated task ultimately leads to e.g. the flow modelling of the CVD reactor as done for example by the Flow Expert system of Pons *et al.* [89Pon]. The procedure of modeling a CVD reactor by coupling the thermochemical and kinetic parameters of the process, has been summarised Bernard [90Ber] who also gave further references for more detailed reviews on the subject.

Conclusions

The thermodynamic simulation of a CVD process, based on an appropriate selection of the thermochemical data of the participating species, helps to avoid a number of general errors, to save considerable time in finalising the experimental setup and to interpret certain phenomena observed during the experimental procedure. Associated to the flow modeling of the CVD reactor it permits a close approach and consequently the control of the deposition rate, the film uniformity and the composition, i.e. the reactor performance.

References

75Ber C.Bernard, Y.Deniel, A.Jacquot P. Vay and M.J.Ducarroir.: *Less Common Metals* **40**, 1975, 165.

76Duc J.P.Duchemin: Thesis, University of Caen, France, 1976.

76Eri G.Eriksson: *Chemica Scripta* **8**, 1976, 100.

77Bes T.M.Besmann and K.E.Spear: *J. Electroc. Soc.* **124**, 1977, 790.

86Bar J.N.Barbier and C.Bernard: Proceedings of the 15th Calphad Meeting, L.Kaufman, ed., *Calphad*, 1986, 206.

86Hod S.M.Hodson and R.H.Davies: Proceedings of the 15th Calphad Meeting, L.Kaufman, ed., *Calphad*, 1986, 208.

87Ber C.Bernard, C.Vahlas, J.F.Million-Brodaz,J.F and R.Madar: Proceedings of the 10th Intern. Conf. on CVD, Honolulu, USA, G.W.Cullen, ed., *Proc. Electrochem. Soc.*, **87–8**, 1987, 700.

89Pon M.Pons, R.Klein, C.Arena and S.Mariaux: Proc. of the 7th Europ. Conf. on CVD, *J. de Physique*, Colloque C5, suppl. au no.5, 50, 1989, 57.

89Vah C.Vahlas, P.Y.Chevalier and E.Blanquet: *Calphad* **13**(3), 1989, 273.

90Ber C.Bernard: *High Temp. Sc.* **27**, 1990, 131.

91Mas E.Mastromatteo: Thesis, INP Grenoble, France, 1991.

91Tho N.Thomas, E.Blanquet, C.Vahlas, C.Bernard and R.Madar: *MRS Symposium Proceedings* **204**, 1991, 451.

93Tho N.Thomas, J.N.Barbier, C.Bernard, C.Vahlas and R.Madar: *Journal of Chem. Vap. Depos.* **1(3)**, 1993, 263.

13 Calculation of the Concentration of Iron and Copper Ions in Aqueous Sulphuric Acid Solutions as a Function of the Electrode Potential

JÜRGEN KORB* AND KLAUS HACK*

Introduction

When reprocessing metal containing aqueous solutions by electrolysis, it is very useful to know the behaviour and theoretically possible concentrations of the metal ions in the area near the electrodes as a function of the prevailing electrode potential and pH value. The present example concerns a sulphuric acid solution with variable contents of iron and copper, and an acid concentration of 100 g of free H_2SO_4 per litre.

In the first part of this discussion the system is treated with the usual reaction thermochemistry for which the following simplifications are made:

- Because of kinetic inhibitions it is not possible to reduce SO_4/2- or HSO_4/2- ions to elementary sulphur, or HS/- or S/2- ions to H_2S at the cathode. The system will, therefore, be described with SO_4/2- and HSO_4/- as the stable sulphur containing species.
- The total metal content for the elements (Fe,Cu) is calculated as the sum of Fe/2+ and Fe/3+, and Cu/+ and Cu/2+ respectively.
- Because of the lack of higher order interaction terms ideal aqueous behaviour is assumed.
- Neither cathode nor anode materials take part in the reactions.

In the second part, the system is investigated with complex equilibrium methods. The differences to the stoichiometric reaction approach are pointed out.

Since all reactions take place in an aqueous solution it is necessary to take into account the stability range of water when choosing the electrode potential, *Eh*. For O_2 partial pressures of 1 bar the upper limit of stability is described by:

2 H_2O(l) <——> O_2(g) + 4 H/+<aq> + 4 e/-<aq> (13.1)

with $Eh = 1.23 - 0.0592$ pH

* GTT-Technologies, D–52134 Herzogenrath, Germany

The lower limit for variable partial pressure of H_2 is described by:

$$H_2(g) \longleftrightarrow 2\,H/+<aq> + 2e/-<aq>$$

$$\text{with } Eh = -0.0296 \log p_{H2} - 0.0592\ \text{pH} \qquad (13.2)$$

However under practical conditions a wider range of stability of water is found. Excess voltages of more than 0.5 V are generally necessary for a dissociation of the water molecules [65Gar]. For the excess voltage η, which is found for H_2-separation, an empirical equation was given by Tafel [83Paw].

$$\eta = a + 0.117 \log i \qquad (13.3)$$

where, a is a material constant related to the electrode material, and i is the current density. For copper, excess voltages of 0.6 to 0.8 V are found for cathodic current densities of 0.01 to 0.1 A cm^{-2}. With decreasing current density the value of the excess voltage also drops.

In the production of metals by electrolysis of aqueous metal salt solutions lead anodes are often used. For these, one finds for comparable current densities excess voltages of 0.7 V for the oxygen separation.

Considering the above empirical aspects, the range of stability of water is assumed to be between –0.8 V and +2.0 V for the following discussions.

Fe–H_2SO_4–H_2O subsystem

The electrochemical equilibrium between Fe/2+ and Fe/3+ ions in a sulphuric acid solution is discussed for total contents of 1 g l^{-1} (1.7906×10^{-2} mol l^{-1}) and 5 g l^{-1} (8.9503 mol l^{-1}), respectively. Referred to the standard hydrogen electrode, one obtains for the standard electrode potential E° of the half cell Fe/2+<aq> <——> Fe/3+<aq> + e/-<aq> a value of $E^\circ = 0.771$ V.

Under the assumptions outlined in the introduction, the Nernst equation can be employed for the calculation of the concentrations of Fe/2+<aq> and Fe/3+<aq>. One obtains:

$$Eh = E^\circ + RT/F \ln \left(c_{Fe/3+<aq>} / c_{Fe/2+<aq>}\right) \qquad (13.4)$$

$$c_{Fe\_total} = c_{Fe/2+<aq>} + c_{Fe/3+<aq>} \qquad (13.5)$$

$$E1 = (Eh - 0.771)/0.0592 \qquad (13.6)$$

$$c_{Fe/2+<aq>} = c_{Fe\_total} / (1 + 10^{E1})\ \text{mol}\ l^{-1} \qquad (13.7)$$

$$c_{Fe/3+<aq>} = c_{Fe\_total} / (1 + 10^{-E1})\ \text{mol}\ l^{-1} \qquad (13.8)$$

Table 13.1 Concentrations of Fe/2+ and Fe/3+ as a function of the electrode potential for $T = 25°C$ and $c_{Fe\_total} = 1\ g\ l^{-1}$.

Potential	Concentrations	
Eh [V]	Fe/2+ [g l-1]	Fe/3+ [g l-1]
+2.000	1.7373E-21	1.0000E+00
+1.500	4.8508E-13	1.0000E+00
+1.000	1.3542E-04	9.9986E-01
+0.900	6.5778E-03	9.9342E-01
+0.800	2.4454E-01	7.5546E-01
+0.700	9.4056E-01	5.9438E-02
+0.600	9.9871E-01	1.2910E-03
+0.500	9.9997E-01	2.6441E-05
+0.400	1.0000E+00	5.4089E-07
+0.300	1.0000E+00	1.1064E-08
+0.200	1.0000E+00	2.2632E-10
+0.100	1.0000E+00	4.6296E-12
+0.000	1.0000E+00	9.4700E-14
-0.100	1.0000E+00	1.9371E-15
-0.200	1.0000E+00	3.9625E-17
-0.300	1.0000E+00	8.1056E-19
-0.400	1.0000E+00	1.6580E-20
-0.500	1.0000E+00	3.3916E-22
-0.600	1.0000E+00	6.9377E-24
-0.700	1.0000E+00	1.4191E-25
-0.800	1.0000E+00	2.9029E-27

The value of 55.847 g mol^{-1} for the molar mass of iron was used for the conversion of mol l^{-1} into g l^{-1}.

Tables 13.1 and 13.2 contain the calculated concentrations for a total iron content of 1 and 5 g l^{-1} respectively and for a voltage range from +2.0 V to –0.8 V.

The concentration curves given in Figures 13.1 and 13.2 intersect at the equi-concentration points of Fe/2+<aq> and Fe/3+<aq>. From the Nernst equation, a potential of 0.771 V independent of the total iron content is obtained for the chosen conditions and with $Eh = E°$.

Cu – H_2SO_4 – H_2O subsystem

The electrochemical equilibrium between Cu/+ and Cu/2+ ions in a sulphuric acid solution will be discussed for total contents of 1 g l^{-1} ($1.5738.10^{-2}$ mol l^{-1}) and 5 g l^{-1} (7.8691 10–2 mol l^{-1}). Referred to the standard hydrogen electrode one obtains for the standard electrode potential $E°$ of the half cell Cu/+<aq> <——> Cu/2+<aq> + e/–<aq> a value of $E° = 0.167$ V.

Again, accepting the assumptions outlined in the introduction the Nernst

Table 13.2 Concentrations of Fe/2+ and Fe/3+ as a function of the electrode potential for $T = 25°C$ and $c_{Fe\_total} = 5 g\, l^{-1}$.

Potential	Concentrations	
Eh [V]	Fe/2+ [g l^{-1}]	Fe/3+ [g l^{-1}]
+2.000	8.6863E-21	5.0000E+00
+1.500	2.4254E-12	5.0000E+00
+1.000	6.7712E-04	4.9993E+00
+0.900	3.2889E-02	4.9671E+00
+0.800	1.2227E+00	3.7773E+00
+0.700	4.7028E+00	2.9719E-01
+0.600	4.9935E+00	6.4550E-03
+0.500	4.9999E+00	1.3221E-04
+0.400	5.0000E+00	2.7044E-06
+0.300	5.0000E+00	5.5321E-08
+0.200	5.0000E+00	1.1316E-09
+0.100	5.0000E+00	2.3148E-11
+0.000	5.0000E+00	4.7350E-13
-0.100	5.0000E+00	9.6857E-15
-0.200	5.0000E+00	1.9813E-16
-0.300	5.0000E+00	4.0528E-18
-0.400	5.0000E+00	8.2902E-20
-0.500	5.0000E+00	1.6958E-21
-0.600	5.0000E+00	3.4688E-23
-0.700	5.0000E+00	7.0957E-25
-0.800	5.0000E+00	1.4515E-26

equation can be employed for the calculation of the concentrations of Cu/+<aq> and Cu/2+<aq>. One obtains:

$$Eh = E^{\circ} + RT/F \ln \left(c_{Cu/2+<aq>} \,/\, c_{Cu/+<aq>}\right) \quad (13.9)$$

$$c_{Cu\_total} = c_{Cu/+<aq>} + c_{Cu/2+<aq>} \quad (13.10)$$

$$E2 = (Eh - 0.167)/0.0592 \quad (13.11)$$

$$c_{Cu/+<aq>} = c_{Cu\_total} \,/\, (1 + 10^{E2})\ \mathrm{mol}\ l^{-1} \quad (13.12)$$

$$c_{Cu/2+<aq>} = c_{Cu\_total} \,/\, (1 + 10^{-E2})\ \mathrm{mol}\ l^{-1} \quad (13.13)$$

The value of 63.540 g mol^{-1} for the molar mass of copper was used for the conversion of mol l^{-1} into g l^{-1}.

Tables 13.3 and 13.4 contain the calculated concentrations for a total copper content of 1 and 5 g l^{-1} respectively and for a voltage range from +2.0 V to –0.8 V. Figures 13.3 and 13.4 show the appropriate graphical representations.

The values for E° of iron and copper have been taken from the tables of Rauscher–Voigt–Wilke–Wilke [65Rau].

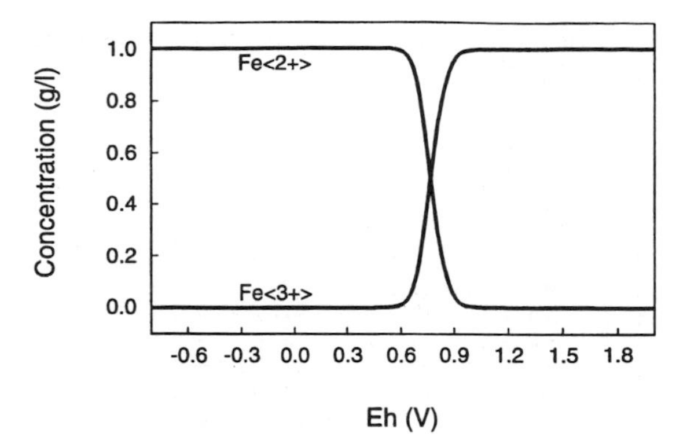

Fig. 13.1 Fe<2+> and Fe<3+> concentration as function of *Eh* at 25°C and 1 g l^{-1} total ferreous content in a sulphuric acid solution.

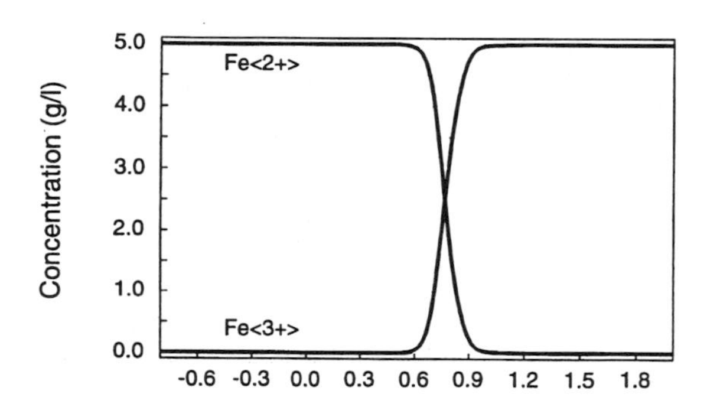

Fig. 13.2 Fe<2+> and Fe<3+> concentration as function of *Eh* at 25°C and 5 g l^{-1} total ferreous content in a sulphuric acid solution.

The complete system Cu–Fe–H_2SO_4–H_2O

In contrast to the 'classical' stoichiometric reaction approach described above, the complete system copper–iron–sulphuric acid–water has been analysed by complex equilibrium computations using the program ChemSage [90Eri]. A system with the phases and species listed in Table 13.5 has been composed from different databases [82Wag, 92SGT].

It should be noted that the last mentioned group of substances in the table is only listed to indicate the completeness of the dataset. None of the solid phases is involved in any of the equilibria discussed in this article.

For the aqueous species, ideal Debye–Hückel behaviour is assumed and the temperature of the system is fixed at 25°C because of a lack of data on activities, i.e. parameters for the non-ideal Pitzer model, and temperature dependence. The composition of the system is defined by the values in Table 13.6 (reference 1 kg = 1 l of water). The total concentration of Cu and Fe respectively is thus 1 g/l each.

Table 13.3 Concentrations of Cu/+ and Cu/2+ as a function of the electrode potential for $T = 25$ °C and $c_{Cu\_total} = 1\ g\ l^{-1}$.

Potential	Concentrations	
Eh [V]	Cu/+ $[g\ l^{-1}]$	Cu/2+ $[g\ l^{-1}]$
+2.000	1.0893E-31	1.0000E+00
+1.500	3.0416E-23	1.0000E+00
+1.000	8.4929E-15	1.0000E+00
+0.900	4.1519E-13	1.0000E+00
+0.800	2.0297E-11	1.0000E+00
+0.700	9.9225E-10	1.0000E+00
+0.600	4.8508E-08	1.0000E+00
+0.500	2.3714E-06	1.0000E+00
+0.400	1.1591E-04	9.9988E-01
+0.300	5.6354E-03	9.9436E-01
+0.200	2.1695E-01	7.8305E-01
+0.100	9.3124E-01	6.8755E-02
+0.000	9.9849E-01	1.5080E-03
-0.100	9.9997E-01	3.0892E-05
-0.200	1.0000E+00	6.3194E-07
-0.300	1.0000E+00	1.2927E-08
-0.400	1.0000E+00	2.6442E-10
-0.500	1.0000E+00	5.4089E-12
-0.600	1.0000E+00	1.1064E-13
-0.700	1.0000E+00	2.2632E-15
-0.800	1.0000E+00	4.6296E-17

Table 13.4 Concentrations of Cu/+ and Cu/2+ as a function of the electrode potential for $T = 25$ °C and $c_{Cu\_total} = 5\ g\ l^{-1}$.

Potential	Concentrations	
Eh [V]	Cu/+ $[g\ l^{-1}]$	Cu/2+ $[g\ l^{-1}]$
+2.000	5.4467E-31	5.0000E+00
+1.500	1.5208E-22	5.0000E+00
+1.000	4.2464E-14	5.0000E+00
+0.900	2.0759E-12	5.0000E+00
+0.800	1.0149E-10	5.0000E+00
+0.700	4.9613E-09	5.0000E+00
+0.600	2.4254E-07	5.0000E+00
+0.500	1.1857E-05	5.0000E+00
+0.400	5.7957E-04	4.9994E+00
+0.300	2.8177E-02	4.9718E+00
+0.200	1.0847E+00	3.9153E+00
+0.100	4.6562E+00	3.4378E-01
+0.000	4.9925E+00	7.5400E-03
-0.100	4.9998E+00	1.5446E-04
-0.200	5.0000E+00	3.1597E-00
-0.300	5.0000E+00	6.4633E-08
-0.400	5.0000E+00	1.3221E-09
-0.500	5.0000E+00	2.7044E-11
-0.600	5.0000E+00	5.5321E-13
-0.700	5.0000E+00	1.1316E-14
-0.800	5.0000E+00	2.3148E-16

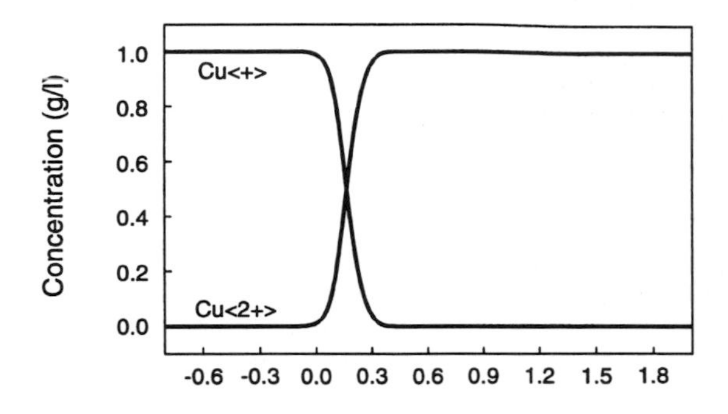

Fig. 13.3 Cu<+> and Cu<2+> concentration as function of *Eh* at 25 °C and 1 g l^{-1} total copper content in a sulphuric acid solution.

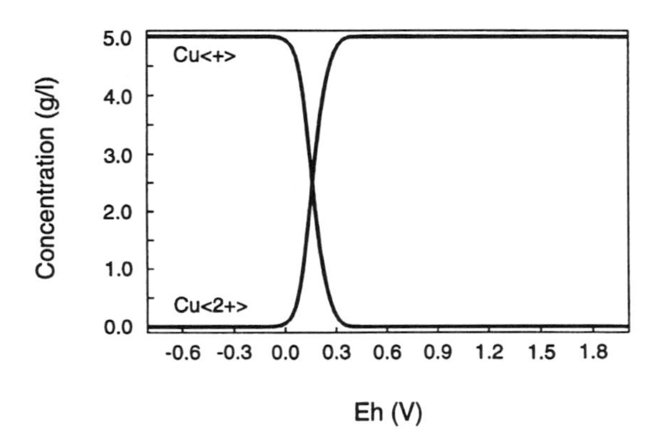

Fig. 13.4 Cu<+> and Cu<2+> concentration as function of *Eh* at 25 °C and 5 g l^{-1} total copper content in a sulphuric acid solution.

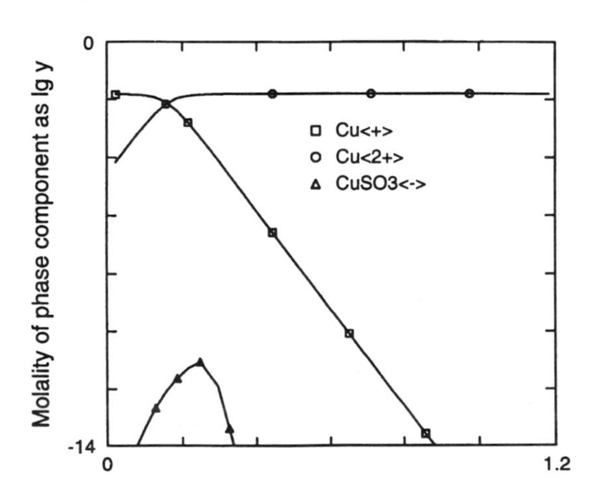

Fig. 13.5 Concentration of Cu species as function of *Eh* at 25 °C and 1 g l^{-1} total copper content in a sulphuric acid solution.

Table 13.5 Composition of the system copper-iron-sulphuric-acid-water.

Gas (g, 4 major species)		
H_2	O_2	SO_2 SO_3
Aqueous solution (aq, 35 species including H_2O):		
H/+	Cu/+	Fe/2+
OH/–	Cu/2+	Fe/3+
SO_2	CuO_2/2–	FeO_2/2–
SO_3/2–	HCuO2/–	FeOH/+
SO_4/2–	$CuSO_3$/–	FeOH/2+
S_2O_3/2–	$Cu(SO_3)_2$/3–	$HFeO_2$/–
S_2O_4/2–	$Cu(SO_3)_3$/5–	$Fe(OH)_2$/+
S_2O_8/2–		$Fe(OH)_3$
S_4O_6/2–		$Fe(OH)_3$/–
HSO_3/–		$Fe(OH)_4$/2–
HSO_4/–		$Fe_2(OH)_2$/4+
H_2SO_3		$FeSO_4$/+
HS_2O_4/–		$Fe(SO_4)_2$/–
$H_2S_2O_4$		
Solid precipitates (25 species (=phases) in total)		
Cu	Fe	S(rhomb.)
CuO	$Fe_{0.947}O$ (Wuest.)	
Cu_2O	Fe_2O_3	
CuS	Fe_3O_4	
Cu_2S	$Fe(OH)_2$	
$CuSO_4$	$Fe(OH)_3$	
$CuSO_4.H_2O$	FeS(alpha)	
$CuSO_4.3H_2O$	FeS_2(pyrite)	
$CuSO_4.5H_2O$	Fe_7S_8	
Cu_2SO_4	$FeSO_4$	
$CuSO_4.2Cu(OH)_2$	$FeSO_4.7H_2O$	
$CuSO_4.3Cu(OH)_2$		
$CuSO_4.3Cu(OH)_2.H_2O$		

The redox potential of all part reactions in the system has the same value at equilibrium. By influencing an arbitrarily chosen pair, e.g. H_2(g)–H/+<aq>, the computational procedure can find the equilibrium as a function of the potential *Eh*.

Figure 13.5 shows the calculated results. Of the copper species considered only Cu/+<aq>, Cu/2+<aq> and CuSO3/–<aq> are of importance up to a molality of $1..10^{-14}$. As expected from the calculations discussed above (Cu – H_2SO_4 – H_2O subsystem) the concentration curves for Cu/+<aq> and Cu/2+<aq> intersect at 0.167 V and $7.869.10^{-3}$ mol Cu_total/l, the logarithm of which is –2.1242. These results are equivalent to those calculated from the Nernst equation (Cu–H_2SO_4–H_2O subsystem).

For iron, however, the major species at equilibrium are Fe/2+<aq>,

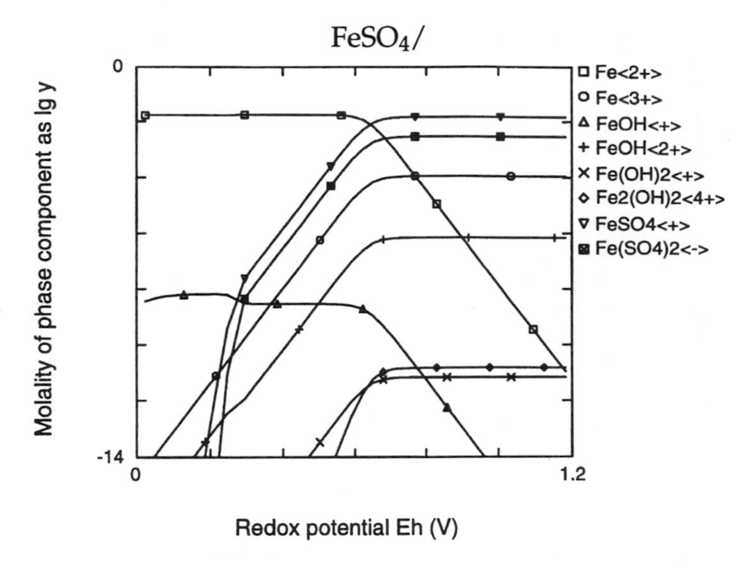

Fig. 13.6 Concentration of Fe species as function of *Eh* at 25 °C and 1 g l^{-1} total ferreous content in a sulphuric acid solution.

Table 13.6 Quantitative composition of the system.

Species	Amount	
H_2SO_4	g l^{-1}	mol l^{-1}
H_2SO_4	100	1.0196
$CuSO_4$	2.52	1.5738×10^{-2}
$FeSO_4$	2.72	1.7906×10^{-2}

Fe/3+<aq>and $Fe(SO_4)_2$/–<aq>. This indicates a major change of the ions that influence the potential in comparison to the choice taken for the calculations with the Nernst equation (Fe – H_2SO_4 – H_2O subsystem). Figure 13.6 shows a decrease of species containing divalent iron (as Fe/2+ and FeOH/+) and an increase of species containing trivalent iron.

As can be expected, the sulphur species HSO_4/–<aq> and SO_4/2–<aq> dominate the aqueous solution from a potential of +0.3 V (see Figure 13.7). The formation of H_2S, solid sulphur (S) and copper- or iron-sulfides does not occur because of strong kinetic inhibitions. Therefore, the hydrogen which is added for the fixation of the potential leads to a reduction of sulphate and formation of S_2O_3/2–<aq> and S_4O_6/2–<aq> ions at potentials less than 0.3 V.

Conclusions and further developments

The calculational results permit an estimation and judgement of the

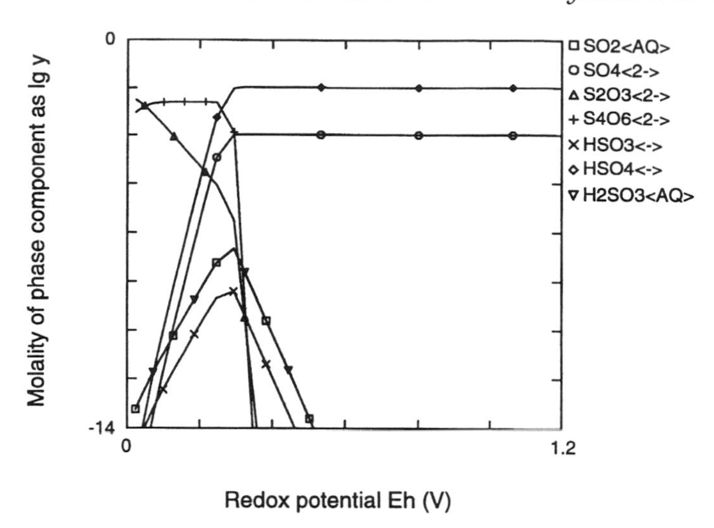

Fig. 13.7 Concentration of S species as function of *Eh* at 25 °C and 1 g l^{-1} total copper and ferreous content in a sulphuric acid solution.

conditions that have to be chosen in potential aqueous metal recovery processes. However, it is evident that the application of the 'classical' method, i.e. the exclusive use of the Nernst equation for a preselected stoichiometric reaction, permits no comprehensive understanding of the equilibrium state of a complex multicomponent system. For copper, the application of the simple method was successful, because the assumed reaction was the correct one. For iron, it was shown that the initial choice, reasonable though it seemed, was not good. The formation of complexes of multivalent elements can lead to a considerable shift in the reactions, that actually define the potential. Complex equilibrium computation has been used successfully to overcome this problem.

It should, however, be noted that more advanced calculations would have to include non-ideal interactions between the aqueous species since the compositional range covered in the present calculations exceeds the validity range of the Debye–Hückel limiting law. Furthermore, the present results only pertain to 25 °C. If higher temperatures are of interest, C_P functions of the aqueous species have to be introduced. Indeed, in some cases, it is even necessary to derive values for H_{298} and S_{298}, since only ΔG_{298} are presently available for many aqueous species. Experience shows that extrapolation of such simplified room temperature data beyond 35°C can lead to unacceptably large deviations.

The calculations for this case study have been performed using dedicated software and the program ChemSage.

References

65Gar R.M.Garrels and Ch.L.Christ: *Solutions, Minerals and Equilibria,* A Harper international Student Reprint, published by Harper & Row, New York, Evanston & London and John Weatherhill, Inc., Tokyo, 1965.

65Rau K.Rauscher, J.Voigt, I.Wilke, and K.-Th.Wilke: *Chemische Tabellen und Rechentafeln für die analytische Praxis,* VEB Deutscher Verlag für Grundstoffindustrie, Leipzig, 1965.

82Wag *The NBS Tables of Chemical Thermodynamics Properties, Selected Values for Inorganic and C1 and C2 Organic Substances in SI Units,* American Chemical Society and the American Institute of Physics for National Bureau of Standards.
Wagman,D.D.; Evans,W.H.; u.a.: Journal of Physical and Chemical Reference Data **11**, 1982, Supplement No. 2.

83Paw F.Pawlek: *Metallhüttenkunde,* Walter de Gruyter, Berlin, New York, 1983.

90Eri G.Eriksson and K.Hack: *Met. Trans. B* **21B**, 1990, 1013–1024.

92SGT *SGTE Pure Substance Database,* 1992 edition.

14 Thermochemical Conditions for the Production of Low Carbon Stainless Steels

KLAUS HACK*

Introduction

Austenitic stainless steels classically contain the elements iron, chromium and nickel; chromium (typically 18 wt%) to give corrosion resistance and nickel (typically 8 wt%) to improve the ductility of the Fe-Cr-alloys. However, during the production of these alloys, the presence of carbon can never be avoided. Unfortunately, higher carbon contents lead to the formation of $M_{23}C_6$ carbides mainly containing chromium, thereby reducing the corrosion resistance. It is, therefore, necessary to establish a production process that yields Fe-Cr-Ni-C alloys with low carbon and high chromium contents. In the converter, oxygen is blown through the liquid Fe-bath, thus removing carbon as CO gas. However, chromium also has a high affinity for oxygen, and oxidation of chromium from the melt is quite probable, especially if the activity of chromium oxide is reduced by it being dissolved in the slag.

Thus, the thermochemical problem is one of a six component system, the components being:

Fe, C, Cr, Ni, O and possibly Ar (see below).

The major phases to be considered are: metallic melt, gas phase and a slag (of unknown composition).

The temperature and pressure conditions are as yet undefined. This situation is summarised in Table 14.1.

The mass action law approach

As carbon and chromium both compete for the oxygen the following stoichiometric exchange reaction can be used to describe the decisive equilibrium:

$$2\,Cr + 3\,CO\langle G\rangle = 3\,C + Cr_2O_3 \qquad (14.1)$$

However, the reactants are not pure substances. Instead, chromium and carbon will be dissolved species in an iron-based melt, Cr_2O_3 can be dissolved

* GTT-Technologies, D–52134 Herzogenrath, Germany.

Table 14.1 Summary of the thermochemical aspects of the production of low carbon stainless steel.

Global conditions	wt% Ni = 8
Elements	Fe, C, Cr, Ni, O, (Ar)
Phases	Liquid metal, Gas, Slag

in a slag or precipitate as pure solid oxide and CO is a species in the gas phase. Thus, none of the activities of the reactants or products is equal to unity. Temperature and total pressure in the process are variables, which in practice can be modified over a wide range. The mass action law provides a relationship between all the parameters to decide which of them is the most important and should, therefore, be controlled in practice. One obtains from the equilibrium condition:

$$\log a_C = {}^2/_3 \log a_{Cr} + {}^1/_3 (\log K - \log a_{Cr2O3} + 3 \log p_{CO}) \quad (14.2)$$

For the activity of chromium, a value near 0.2 can be assumed since the iron–chromium system is nearly ideal and an 18/8 steel contains about 20at.% of chromium. The temperature and the partial pressure of CO can be controlled in a converter and, to a lesser extent, the activity, i.e. the contents, of Cr_2O_3 in the slag. In Figure 14.1 the phase diagram of the slag system CaO–SiO_2–Cr_2O_3 is shown. The slag path in a converter is overlaid, showing that the Cr_2O_3 amounts, and thus the activities, have essentially very low values. Only in the intermediate stage of the process might precipitation of Cr_2O_3 (i.e.a_{Cr2O3}=1) be reached. The losses of chromium to the slag are usually recovered, as can be seen from last section of the slag path. For a quick check at a reasonable value of the Cr_2O_3 activity (a_{Cr2O3}=0.1) a matrix (Table 14.2) of the carbon activity as a function of temperature and CO–partial pressure can be obtained.

Table 14.2 Carbon activity as a function of temperature and CO partial pressure for $a_{Cr2O3} = 0.1$ and $a_{Cr} = 0.2$.

P_{CO}/bar	T/K		
	1800	1900	2000
1	2.9E-2	1.2E-2	5.2E-2
0.1	2.9E-3	1.2E-3	5.2E-3
0.01	2.9E-4	1.2E-4	5.2E-4

The trends in the lines and columns show that both parameter changes (increase in temperature and decrease in P_{CO}) result in a decrease of the carbon activity, i.e. the carbon content in the iron bath. However, a change in P_{CO} has a far greater effect than a change in temperature.

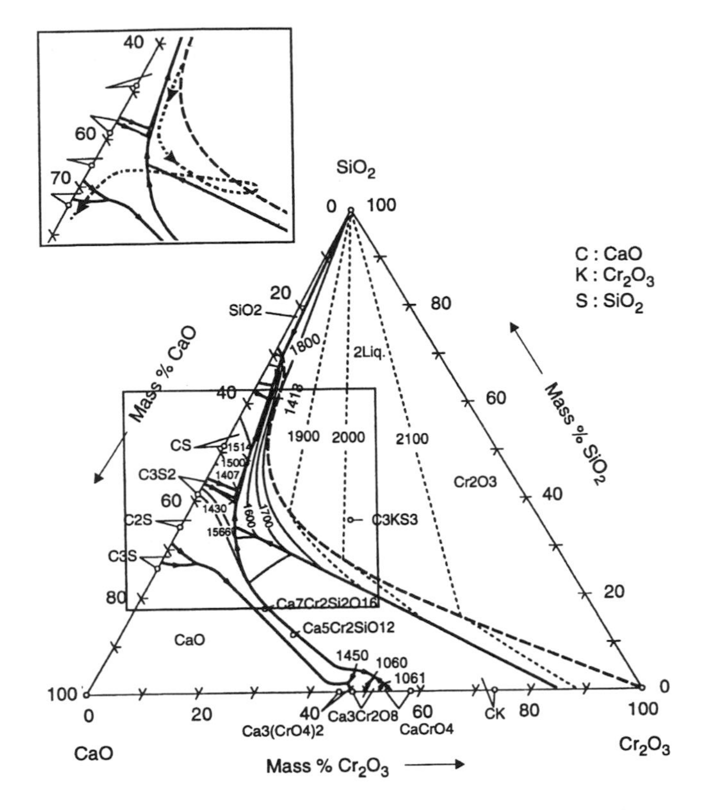

Fig. 14.1 Phase diagram of the slag system CaO–SiO_2–Cr_2O_3 [83Gei, 92Kow]. Insert: a typical slag path.

The complex equilibrium approach

If confirmation is required that a certain level of carbon content (e.g. 0.01 wt%) is reached, only a series of proper complex equilibrium calculations will provide the desired result [90Spe].

Figure 14.2 shows wt% C vs wt% Cr for different temperatures between 1600 and 1900 °C. The concentration of Ni is set to be 8 wt%, the value for p_{CO} is set to 0.1 atm, and the activity of Cr_2O_3 is fixed at 0.1. All values are representative for practical conditions. From the diagram, it is obvious that increasing temperature reduces the concentration of carbon; however, with detrimental consequences for the refractories in the converter and at high energy costs.

Figure 14.3 shows the same relationship for different values of the activity of Cr_2O_3. The temperature is fixed to an intermediate value of 1700 °C, the partial pressure of CO and the Ni concentration are kept as in Figure 14.2 (0.1 atm and 8 wt%). Clearly, only increasing the activity of Cr_2O_3 to a value of one will reduce the concentration of carbon to the desired level, resulting in the transfer of chromium from the melt to the slag (See Figure 14.1).

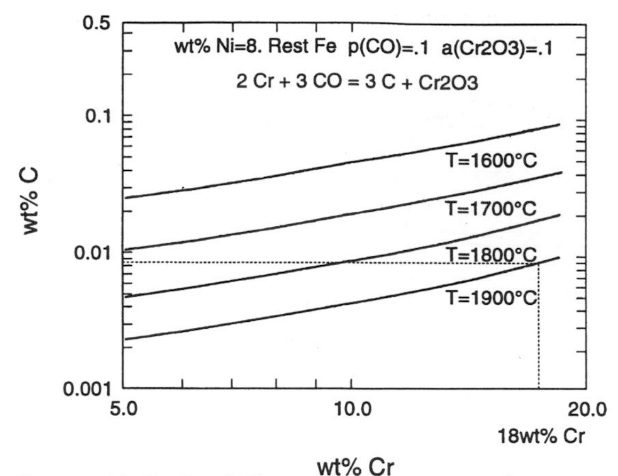

Fig. 14.2 wt% C vs wt% Cr for different temperatures between 1600 and 1900 °C.

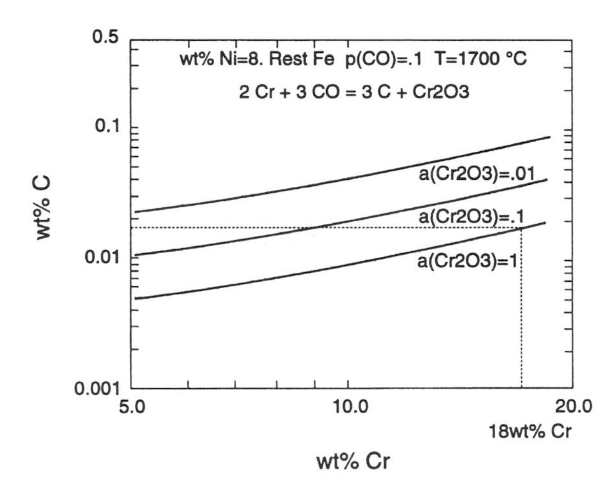

Fig. 14.3 wt% C vs wt% Cr for different values of the activity of Cr_2O_3.

Figure 14.4 shows the influence of a variation of p_{CO} at a fixed activity of Cr_2O_3 (0.1), fixed concentration of Ni (8 wt%) and again for T=1700°C. This diagram shows the strongest trend of the curves with variation of the parameter.

Engineering conclusions

It is obvious that of all the parameters varied, the partial pressure of CO is the most important. The lower its value (with all other parameters fixed), the lower the carbon concentration in the iron bath.

The engineering solution can now be either reduction of the total gas pressure level, a process that is called Vacuum Oxygen Decarburisation (VOD), or reduction of the partial pressure of CO by strong dilution in the gas phase, called Argon Oxygen Decarburisation (AOD) [87Lin].

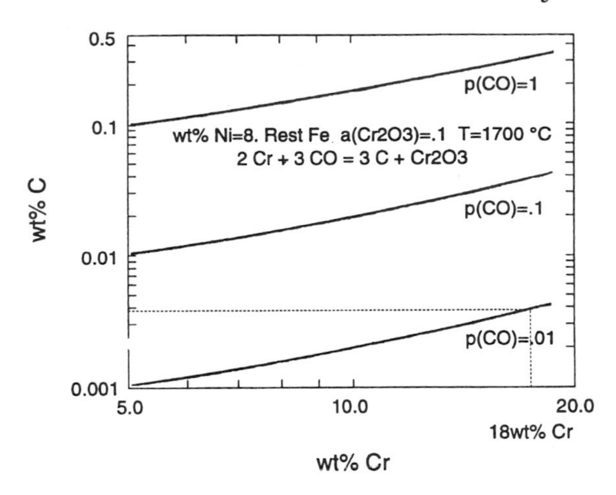

Fig. 14.4 wt% C vs wt% Cr with varying p_{CO} at a fixed activity of Cr_2O_3 (0.1), fixed concentration of Ni (8 wt%) and for *T*=1700 °C.

It should be noted that using calculations based on the stoichiometric reaction, e.g. the law of mass–action including Gibbs energy data for pure substances, the assumptions about the possible values of the equilibrium activities already enable these conclusions to be made. In principle, a thermochemist of the thirties would have been able to perform these calculations since all the Gibbs energy data needed were available at that time. Although the majority of this discussion is qualitative in nature, the results show that the conclusion as to which is the major parameter in the process is stringent.

However, in the calculations above, it was possible to tackle the quantitative aspects of the redox reaction only because the Gibbs energy models and data available today permit complex equilibrium calculations to take into proper account the concentrations of the different components in the different solution phases. For example, it is now possible to determine if a defined pressure level is sufficient to keep the concentration of C below 0.01 wt%, the limit of formation of $M_{23}C_6$. The transformation of carbon and chromium activities into concentrations can be carried out for liquid steels taking into account the presence of 8 wt% nickel. With thermodynamic data available for the Gibbs energy of mixing of multicomponent non-ideal metallic Fe melt as well as for a multicomponent oxidic slag, it is possible to produce the above series of diagrams. Thus, the question concerning the low carbon concentration can also be answered, but only using complex equilibrium calculations.

Acknowledgements
The author wishes to thank Prof. D. Neuschütz, LTH, RWTH Aachen, for his helpful discussions. The calculations for this case study have been performed using ChemSage.

References

83Gei J.GEISELER, K.GRADE, and P.VALENTIN: *Stahl und Eisen* **103**, 1983, 1013-1017.

87Lin H.-U.LINDENBERG, K.-H.SCHUBERT and Z.ZÖRCHER: *Stahl und Eisen* **107**, 1987, 1197-1204.

90Spe P.J.SPENCER and K.HACK: *Swiss Materials* **2,3a**, 1990, 69.

92Kow M.KOWALSKI, P.J.SPENCER and D.NEUSCHÜTZ: *Evaluation and critical compilation of thermochemical data and physical property values of slag for iron and steelmaking, phase diagrams*, part 3, ECSC research report 7210-CF/107, 1992.

15 Application of Phase Equilibrium Calculations to the Analysis of Severe Accidents in Nuclear Reactors

RICHARD G. J. BALL*, PAUL K. MASON** AND MIKE A. MIGNANELLI**

Introduction

During the design and construction of a nuclear reactor many safety features are built into the plant to ensure the reactor can be operated safely or shut down following the occurrence of a fault within the design basis. However, during a severe accident, large scale degradation of the core could occur. Such faults are beyond the design basis and although they are extremely improbable, the UK Nuclear Installations Inspectorate (the licensing body) requires a demonstration that there is no sudden unacceptable increase in the consequences of accidents just beyond the design basis. To assess the consequences of a severe accident requires the systematic analysis and quantification of releases of radioactivity and the consideration of all possible processes which could significantly contribute to the release (source term). Research efforts are, therefore, focused at understanding the physical and chemical phenomena relevant to severe accidents, with a view to providing a sufficient scientific basis to permit realistic source terms to be established. Severe accident analyses are most extensively developed for Light Water Reactors (LWRs) and, in the UK, most interest has been placed the Pressurised Water Reactors (PWRs). The work described in this chapter is, therefore, based on the PWR.

A schematic diagram of a typical PWR is shown in Figure 15.1. The cylindrical pressure vessel is constructed of steel and contains the reactor core and control rods, with the remaining volume occupied by ordinary 'light' water under a pressure of about 150 atmospheres. The water acts as both the moderator and primary coolant. The core is made up of a number of fuel elements which are a collection of fuel pins comprising Zircaloy-4 alloy cladding enclosing uranium dioxide (UO_2) fuel. As shown in Figure 15.1, the reactor pressure vessel is surrounded by two or more metres of concrete which is designed to form a secondary containment should the reactor pressure vessel fail during a fault. This concrete containment building provides a barrier to the release of fission products into the external atmosphere. In the unlikely event of failure of the secondary containment, fission products could be released into the atmosphere.

*Materials and Chemistry Group, AEA Technology, 1700 N. Highland Road, Pittsburgh, PA 15241, USA
*Chemical Process Assessment, AEA Technology, Harwell, Didcot, Oxfordshire OX11 0RA, UK.

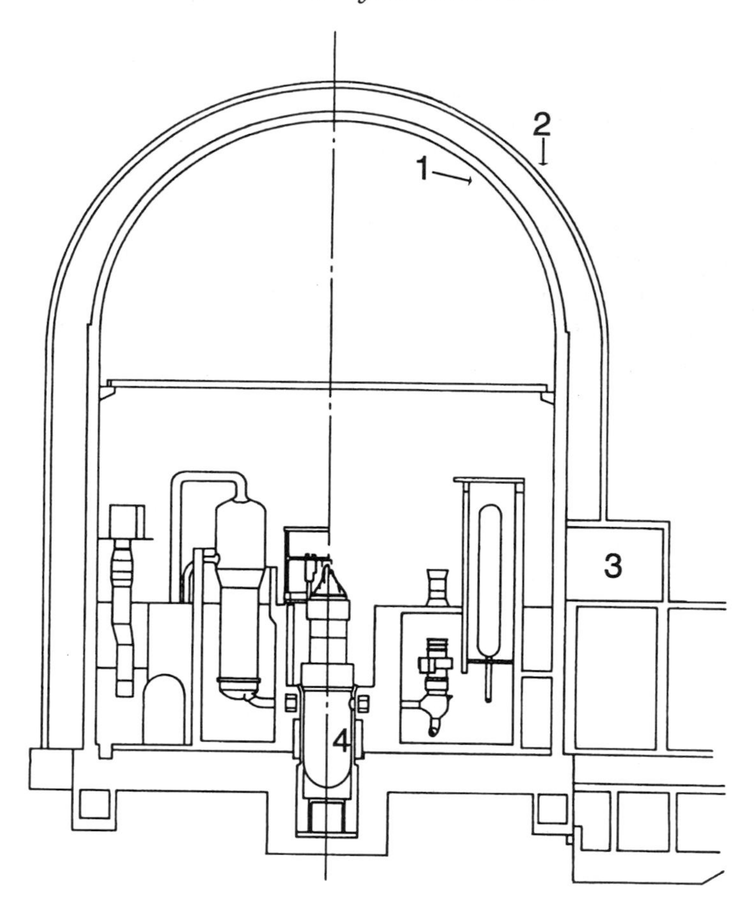

Fig. 15.1 Schematic diagram of a pressurised water reactor. (1) Primary containment; (2) Secondary containment enclosure building; (3) Auxiliary buildings; (4) Reactor pressure vessel.

During the progression of a severe accident, the fuel and cladding could melt and slump to the bottom of the steel pressure vessel. The corium, as this mixture of fuel and cladding is referred, may then melt through the floor of the pressure vessel and be ejected onto the concrete basemat of the secondary containment building. To predict the progression of the fault requires that the thermal hydraulics and chemical interactions at each stage of the accident can be modelled accurately.

Initially, the chemical state of the fuel and fission products in the intact pins must be known. As the temperature of the pin increases and failures occur, the chemical interaction of the fuel with the Zircaloy cladding material and coolant environment has to be considered. During the stage where significant melting of the core takes place and material is relocated into the bottom of the reactor pressure vessel, immiscible oxide and metal liquids are formed which are initially contained within a crucible of condensed material. After the wall of the reactor pressure vessel is breached, the molten corium, comprising components of the core and structural materials, will be ejected onto the concrete basemat of the secondary containment building. The concrete basemat would then be ablated as a result of the thermal interaction and the composition of the phases would change progressively as the concrete decomposition products comprising oxides and gases are subsumed into the melt. These oxide and metal phases are either layered or intimately mixed due to the sparging of the concrete decomposition gases through the melt. The erosion of the concrete due to molten core-concrete interaction (MCCI) is a threat to containment integrity if complete penetration of the basemat is achieved. In addition, fission products, fuel and the components of the core structural materials and concrete could be released from the melt into the cavity atmosphere by vaporisation. The extent of the vaporisation of these species from the MCCI melt will be dependent on the temperature of the melt, the decomposition gases passing through the melt and the thermodynamic activities of the components, in particular the oxygen potential of the system [86Smi, 86But, 92Mig].

Models which describe the phase equilibria in MCCI melts are important in the prediction of both the extent of concrete penetration and the release of species by vaporisation. In the former, the melt solidus and liquidus temperatures that are determined for a specific composition provide input data to heat transfer calculations and, therefore, influence the predictions of the melt temperature and the onset of solidification. In addition, the heat transfer properties of the melt are also controlled by the viscosity of the melt which is dependent on the amounts and composition of the liquid and solid phases. The extent of release of species during MCCI is determined by the predicted temperature of the melt and the chemical activities of the components in the oxide and metal phases.

To model the thermophysical and chemical behaviour of core debris during the progression of MCCIs it is necessary to have a good understanding of the chemical equilibria in the gas phase and multi-component oxidic and metallic solution phases. Such information can be obtained from calculations of the equilibrium composition of the system using thermodynamic models that describe the components and interactions between the components. The calculations can be performed using codes that determine the minimum of the total Gibbs energy of the system. This development of the thermodynamic models requires the optimisation of the available thermodynamic data and phase equilibrium information. The components of the system required to model the important chemical processes for in-vessel events and core-concrete interactions are: fuel (UO_2), Zircaloy cladding ($Zr–ZrO_2$), concrete constituents (SiO_2, CaO, Al_2O_3, MgO), structural steel components ($Fe–O$) and some refractory fission products (BaO, SrO and La_2O_3).

This chapter describes the application of a model for the nine component UO_2–ZrO_2–SiO_2–CaO–MgO–Al_2O_3–SrO–BaO–La_2O_3 oxide system, based on data for the binary and ternary sub-systems. The sub-oxide systems for $U–UO_{2+x}$, $Si–SiO_2$

and Zr-ZrO_2 and the metal-metal systems included in the model are also referred. Using this model, some calculations of the phase constitution of the multi-component, multi-phase oxide system have been performed as a function of composition and temperature. The oxide database has also been extended to include the gas phase species of the ten element system. Using this extension, calculations have been performed to determine the extent of release of the components by vaporisation and the results of these calculations have been compared with those from models which assume that the oxide solutions are treated as ideal.

Thermodynamic model

The calculation of the thermodynamic equilibria for MCCIs requires a representation of the Gibbs energy of the total system as a function of temperature, pressure and composition. The equilibrium state is then obtained by minimising the total Gibbs energy with respect to the composition and the other input parameters. The total Gibbs energy of the system is given by the weighted sums of the Gibbs energies of the individual phases present at equilibrium. For this work, the weighted sum of the enthalpies (*H*) of the constituent elements in their standard state at 298.15 K were used as a reference for the Gibbs energy of that phase. This is called the standard element reference (*SER*) and hence the Gibbs energies are expressed as *G–H(SER)*. The temperature dependence of the Gibbs energy is expressed in the form:

$$G\text{–}H(SER) = A + BT + CT\ln T + DT^2 + ET^3 + FT^{-1} \tag{15.1}$$

and the various parameters are obtained by fitting to experimental enthalpy, entropy and heat capacity data. If the phase is a pure substance then the Gibbs energy can usually be obtained from standard sources of thermodynamic data such as the SGTE (Scientific Group Thermodata Europe) pure substance database [87Ans, 85JAN, 90Cor]. However, in some cases, intermediate phases in the binary and ternary systems have been reported in experimental studies [90Cor] for which no thermodynamic data are available. In these cases, the relevant parameters, therefore, have been estimated and the values optimised with respect to the known data.

The Gibbs energies for solution phases are more complicated to express. An important parameter for a solution phase is the Gibbs energy of mixing, ΔG^{mix}, which is the change in Gibbs energy accompanying the formation of the solution from its constituents. For an ideal solution, ΔG^{mix} is simply given by the temperature multiplied by the configurational entropy change on forming the solution. In a non-ideal solution, however, ΔG^{mix} will be given by the ideal configurational contribution plus an excess Gibbs energy term, G^{xs}. The modelling of multi-component systems is mainly concerned with deriving suitable representations for G^{xs} which reproduce the experimental phase relationships of a particular system. In some cases, this can be achieved by using a polynomial in composition and temperature to describe the binary and higher order interactions between the components. However, for some solution phases, in which there are strong interactions between atoms or molecules of the different components of the solution and, therefore, pronounced minima in the excess Gibbs energy as a function of composition, a different representation is needed. The representation used in this study for the liquid phase is an associated solution model [82Som] in which one or more associates A_pB_q, are included in the description of the solution between species A and B. The introduction of the associates does not change the form of the Redlich–Kister expression but the values of x_i, which includes the amount of the associate, become variables. For some solid solution phases, a sublattice model [85Hil] has been employed in which mixing occurs on individual sublattices.

In the development of the representations for the components of the chemical model for MCCIs, much use has been made of programs which optimise the coefficients of the Gibbs energy equations to best describe the solution phases [77Luk]. The process involves the assessment of all the available data for all the combinations of the system components. Although, only binary and ternary data have been critically assessed, a multi-component solution phase can usually be adequately modelled using the data from the lower-order systems [87Bha]. However, an important requirement is that assessments of the individual sub-systems are interconsistent; for example, in the assessments of the binary systems, the same data must be used for the pure components in each phase and common solution models must be used. Having derived representations for the Gibbs energies of all the phases in a system, the equilibrium state can then be determined by minimising the total Gibbs energy using a suitable equilibrium code.

Development of the MCCI model

During the early phases of an MCCI, the melt would comprise both oxide and metal immiscible solutions. As described above, the oxide solution would contain the fuel and some of the core structures which have become oxidised. As well as the constituents of the core structures, the metal solution would also contain metal components from the fuel, fission product and concrete (formed by the reduction of the component oxides with zirconium). The influence of the progressive oxidation of the metal phase, due to the sparging CO_2 and H_2O gases, on the composition of the melt is also an important factor in the modelling. Therefore, in addition to the data for the oxide components, relevant data for the metal and sub-oxide phases are also required.

The development [93Bal, 92Che] of the nine component (UO_2–ZrO_2–SiO_2–CaO–MgO–Al_2O_3–BaO–SrO–La_2O_3) oxide thermodynamic database to model the behaviour of the important components of core debris during accident progression is a considerable task. To make greater progress, a collaborative programme was established between AEA Technology, National Physical Laboratory (NPL), Thermodata and the Commissariat a l'Energie Atomique (CEA) in which there was agreement between specialists on the unary data and models used in the phase diagram calculations. The $CaO–Al_2O_3–SiO_2$ system was assessed by the NPL and for the $SiO_2–UO_2–ZrO_2$ system, assessments were made of the binary phase diagram data by AEA Technology in conjunction with Thermodata. In addition, Thermodata have assessed the $BaO–(UO_2–ZrO_2–SiO_2–CaO–Al_2O_3)$, $SrO–(UO_2–ZrO_2–SiO_2–CaO–Al_2O_3)$ and SrO–BaO systems, while AEA Technology have introduced the SrO–MgO, BaO–MgO and $La_2O_3–(UO_2–ZrO_2–SiO_2–CaO–Al_2O_3–MgO–BaO–SrO)$ systems into the model. Table 15.1 summarises the phases included in the nine component oxide database.

As emphasised in this section, consideration of the sub-oxide and metal systems is also required to assess the interactions which may occur in the melt. Therefore, assessments of the $U–UO_2$ and $Si–SiO_2$ systems by Rand [93Ran] and the $Zr–ZrO_2$ system by Lukas [92Luk] have been included in the database using data which are consistent with the models in the oxide database. The data for the metal systems have been obtained from the SGTE metal solution database, which together with data assessed for the structural and cladding components of the reactor and for the control rod material, the Ag–In–Cd system [92Hor], comprise an extensive database for the metal phases of MCCI melts.

Calculations performed using the thermodynamic database

The assessed data for the nine component oxide system $UO_2–ZrO_2–SiO_2–CaO–MgO–Al_2O_3–BaO–SrO–La_2O_3$ have been used to estimate the thermodynamic behaviour of typical core-concrete melts and the release of species by vaporisation from the melt. In addition, calculations have been performed to simulate large scale MCCI experiments. The model that has been developed is providing a significant contribution in the development of the codes used to model the physico-chemical behaviour of MCCIs and releases by vaporisation. Described below are some applications of the model.

a) Solidus–liquidus temperatures for compositions appropriate to MCCIs

A number of sensitivity studies have been carried out to estimate the solidus-liquidus surfaces for compositions appropriate to MCCIs. The parameters that have been studied are the influence of the extent of oxidation of the zirconium in the core debris ($Zr : ZrO_2$ ratio) and different concrete types. The phase equilibria for the system $UO_2–ZrO_2–SiO_2–CaO–Al_2O_3$, for compositions ranging from pure corium (comprised of urania and zirconia) to pure concrete, were calculated over the temperature range 1000 to 3200 K.The two dimensional phase maps for the systems comprising 60%oxidised zirconium and siliceous and limestone concrete, calculated for

Table 15.1 Phases included in the oxide model database.

$AlCa_{0.5}O_4Si$<LIQUID>	Al_2O_4Sr<AL2SRO4_B>	Ca_3O_5Si<HATRURITE>
$AlO_{1.5}$<LIQUID>	$Al_2O_6Sr_3$<AL2SR3O6>	$Ca_3O_7Si_2$<RANKINITE>
$Al_{1.333333}O_4Si$<LIQUID>	Al_4O_7Sr<AL4SRO7>	CaO_3Zr<CAZRO3>
BaO<LIQUID>	$Al_{12}O_{19}Sr$<AL12SRO19>	La_2O_5Si
CaO<LIQUID>	$Al_{32}O_{132}Sr_{84}$<AL32SR8	$La_2O_7Si_2$
CaO_3Si<LIQUID>	BaO<FCC_FLUORITE>	$La_4O_{12}Si_3$
Ca_2O_4Si<LIQUID>	CaO<FCC_FLUORITE>	La_4O_7Sr
$LaO_{1.5}$<LIQUID>	$LaO_{1.5}$<FCC_FLUORITE>	$La_4O_9Sr_3$
La_4O_7Sr<LIQUID>	MgO<FCC_FLUORITE>	$La_2O_7Zr_2$
MgO<LIQUID>	OSr<FCC_FLUORITE>	MgO<TETRA>
Mg_2O_4Si<LIQUID>	O_2U<FCC_FLUORITE>	O_2U<TETRA>
O_2Si<LIQUID>	O_2Zr<FCC_FLUORITE>	O_2Zr<TETRA>
OSr<LIQUID>	BaO<LA2O3>	MgO_3Si<CLINO_P>
O_2U<LIQUID>	CaO<LA2O3>	MgO_3Si<ORTHO_P>
O_2Zr<LIQUID>	$LaO_{1.5}$<LA2O3>	MgO_3Si<PROTO_ENST>
Al_2BaO_4<AL_2BAO_4>	OSr<LA2O3>	Mg_2O_4Si<OLIVINE>
$Al_2Ba_3O_6$<AL2BA3O6>	O_2Zr<LA2O3>	MgO_6Si_2Sr
$Al_{12}BaO_{19}$<AL12BAO19	$BaLa_2O_4$<BALA2O4>	MgO_7Si2Sr_2
Al_2CaO_4<C1A1>	BaO_3Si<BASIO3>	$MgO_8Si_2Sr_3$
$Al_2Ca_3O_6$<C3A1>	BaO_5Si_2<BASI2O5>	$MgO_{12}Si_4Sr_3$
Al_4CaO_7<C1A2>	Ba_2O_4Si<BA2SIO4>	O_2Si<BETA_QUARTZ>
$Al_{12}CaO_{19}$<C1A6>	$Ba_2O_8Si_3$<BA2SI3O8>	O_2Si<CRISTOBALITE>
$Al_{14}Ca_{12}O_{33}$<C12A7>	$Ba_3O_{13}Si_5$<BA3SI5O13>	O_2Si<QUARTZ>
$Al_2CaO_8Si_2$<ANORTHITE>	$Ba_5O_{21}Si_8$<BA5SI8O21>	O2Si<TRIDYMITE>
$Al_2Ca_2O_7Si$<MELILITE>	BaO_3U<BAUO3>	O_3SiSr<SISRO3>
$AlLaO_3$	BaO_3Zr<BAZRO3>	O_4SiSr_2<SISR2O4>
$Al_7La_{33}O_{60}$	$CaMgO_4Si$<CAMGSIO4>	O_5SiSr_3<SISR3O5>
$Al_{11}LaO_{18}$	$CaMgO_6Si_2$<CAMGSI2O6>	O4SiZr<ZIRCON>
$Al_4Mg_2O_{18}Si_5$<CORDIERITE>	$Ca_2MgO_7Si_2$<CA2MGSI2O	O_3SrZr<SRZRO3>
Al_2O_3<ALPHA>	$Ca_3MgO_8Si_2$<CA3MGSI2O	O_4Sr_2Zr<SR2ZRO4>
Al2O3<DELTA>	$CaO_2Si_{0.5}$<APRIME°C2S	$O_7Sr_3Zr_2$<SR3ZR2O7>
Al_2O_3<GAMMA>	CaO_3Si<PSEUDO_ WOLLASTONITE>	O_2Zr<MONO>
Al_2O_3<KAPPA>	CaO_3Si<WOLLASTONITE>	
Al_2O_4Sr<AL2SRO4_A>	Ca_2O_4Si<ALPHA_C2S>	

isopleths across the system, are shown in Figure 15.2. The temperature at which solid begins to form from the liquid solution (the liquidus) and the temperature at which the solid solution just begins to liquefy (the solidus) are shown as bold lines on the diagram and can be followed together with other phase changes that occur in the system as the temperature is changed.

(a)

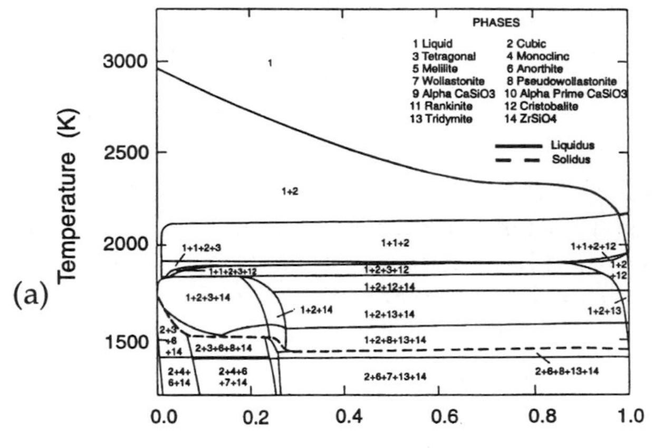

(b)

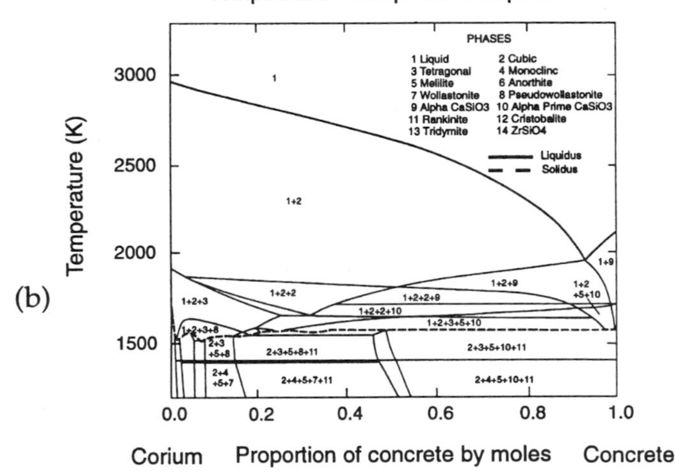

Fig. 15.2 Isopleths from 100% corium to 100% concrete compositions.

(a) *Siliceous concrete*

Composition	Corium/moles	Concrete/moles
UO_2	.730	.0001
ZrO_2	.2700	.0001
SiO_2	.0001	.8670
CaO	.0001	.1150
Al_2O_3	.0001	.0180

b) *Limestone concrete*

Composition	Corium/moles	Concrete/moles
UO_2	.7300	.0001
ZrO_2	.2700	.0001
SiO_2	.0001	.3580
CaO	.0001	.5980
Al_2O_3	.0001	.0440

The importance of the predicted solidus temperatures is greater than that of liquidus temperatures in modelling MCCIs since the former is used to predict other factors. These include the timing and extent of crust formation between the melt-concrete and melt-atmosphere interface which can lead to thermal insulation and so higher melt temperatures. Also the solidus temperature is used as an input parameter in some computer codes such as CORCON [84Col], which model thermal hydraulic progression, to estimate the viscosity of the melt. The development of the oxide database and the calculations reported here have shown that some of these models have been overpredicting the solidus temperatures which as can be seen in Figure 15.2 are close to the solidus temperature of the concrete.

The calculated values of solidus-liquidus temperatures obtained using the oxide database have also been compared with the available experimental data for specific core-concrete compositions. Measurements of the solidus and liquidus temperatures of core-concrete mixtures have been carried out by Roche et al. [93Roc] at Argonne National Laboratory using differential thermal analysis (DTA). A comparison of the experimental data with calculations using the oxide thermodynamic database is shown in Table 15.2. The results show that there is good agreement between the experimental and calculated values for the solidus and liquidus temperatures for mixtures of siliceous concrete with corium. However, the liquidus temperatures for the limestone based concrete differ by up to 300 K. This discrepancy could be due to inadequacies in the thermodynamic modelling, or to difficulties in the determination of the experimental points at such high temperatures for these complex systems. Roche et al have suggested that the discrepancy could arise from the low eutectic temperature for the $CaO–UO_2$ system used in the phase diagram optimisations, which is 95K lower than that measured during their own investigation of the system. Further data are required still to check on these results and to validate fully the model for different compositions.

Table 15.2 Comparison of predicted solidus/liquidus temperatures with experimental measurements.

Corium-concrete system	Solidus temperature (K)		Liquidus temperature (K)	
	Calc.	Expt.	Calc.	Expt.
Core + Siliceous	1434	1400	2395	2549
Core + Limestone	1550	1520	2320	>2723
Core + Lime/sand	1450	1360	2490	>2638

b) Evolution of the different phases during the progression of MCCI

Calculations have also been performed to follow the evolution of the different phases in the corium–concrete system during the progression of an MCCI. In particular, the relative amounts and compositions of the liquid and solid phases can be calculated at a given temperature and core-concrete composition obtained from a snapshot of the progression of the MCCI. The results from such calculations, which were used as part of a code comparison exercise to model some large scale experiments, are typified by the one dimensional phase maps shown in Figure 15.3. The calculations predict that during the course of the MCCI, as the temperature falls over the range 2480 K to 2250 K, the ratio of the amounts of liquid to solid is reduced from 3.9 to1.9. The liquid phase comprises predominantly SiO_2 and UO_2, with the amount of urania decreasing as the temperature is reduced. The composition of the solid solution comprises the cubic UO_2–ZrO_2 phase and remains constant during the progression of the MCCI. These calculations provide important information for modelling the viscosity of the melt, and hence the heat transfer processes, in the thermal hydraulics codes.

c) Calculations of the vapour phase over the oxide system

The thermodynamic dataset has also been extended to include the gas phase species of the components of the oxide system, UO_2–ZrO_2–SiO_2–CaO–MgO–Al_2O_3. Calculations have been performed to determine the extent of release of the components by vaporisation. The results have then been compared with calculations assuming all the solution phases are ideal. A comparison with a model comprising a mixture of the stoichiometric compounds has also been performed.

The results from calculations for the 30% oxidation/limestone concrete system, for compositions involving 20 and 80 mol.% concrete, that is, early and late stages of MCCI respectively, are shown in Figure 15.4. The Figure shows the variation of the partial pressures of the dominant gas phase species with temperature for the ideal and non-ideal cases. The results suggest that there are differences in the predicted partial pressures of the gaseous species SiO, UO_3, UO_2 and Mg in applying the two solution models. In the corium-rich system (early stages of MCCI), the pressures of SiO, UO_3 and Mg are increased by a factor of ~2 using the non-ideal model. Although a similar increase in the Mg pressure was determined for the concrete-rich compositions (later stages of MCCI) using the non-ideal model, the pressures of SiO, UO_3 and UO_2 decreased by factors of ~6, ~2 and ~3 respectively. Differences of up to an order of magnitude were noted for the dominant gaseous species between the results for the solution models and the ideal model. The results confirm that the non-ideal interactions in the melt can have a significant influence on the total amount of species in the vapour phase.

The Gibbs energy model for the oxide phase is complex and requires the use of a sophisticated chemical equilibrium code, such as MTDATA, to calculate the composition of systems involving non-ideal liquid and solid solution phases. The calculations can be time-consuming and hence the complex model is not always suitable for use in safety codes developed to

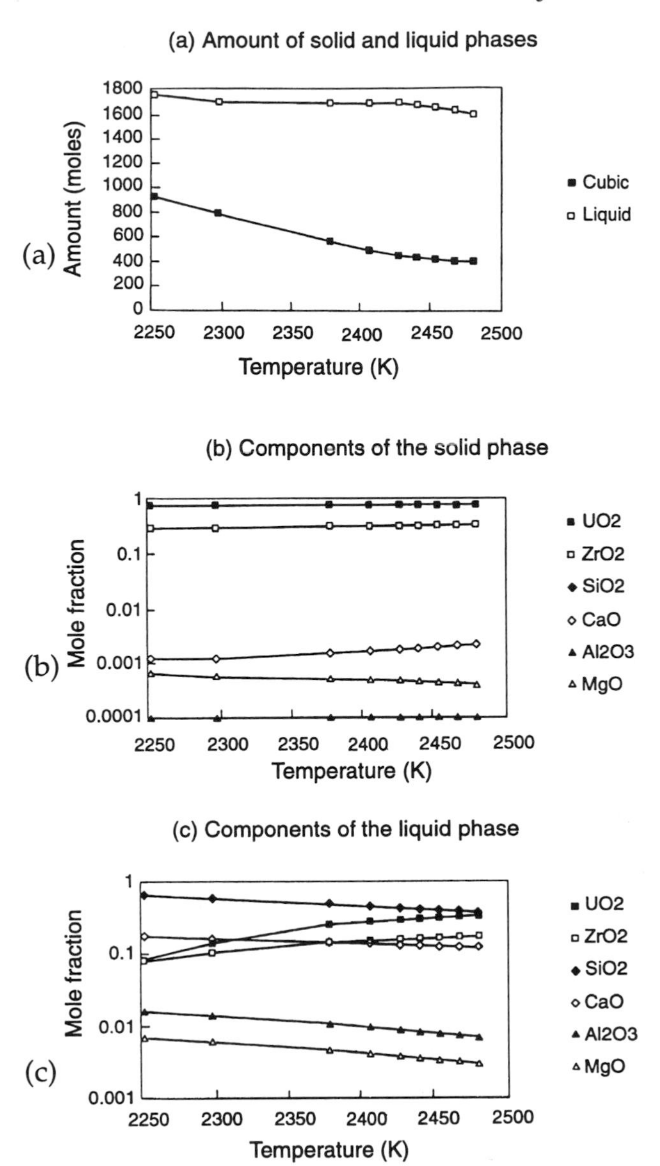

Fig. 15.3 Compositions of the solid and liquid phases of the melt. (a) Amount of solid and liquid phases; (b) Components of the solid phase; (c) Components of the liquid phase.

estimate the thermal hydraulic and chemical behaviour of the melt during the progression of a reactor accident scenario. As many equilibrium calculations are performed it is important that simplified models are used that are fast running in the code but do not add to the uncertainty. Release calculations have been obtained using a combination of the CORCON codeapproach (ideal associated solution). Although the oxide and metal

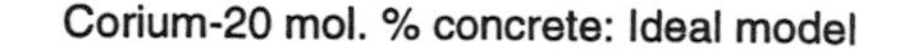

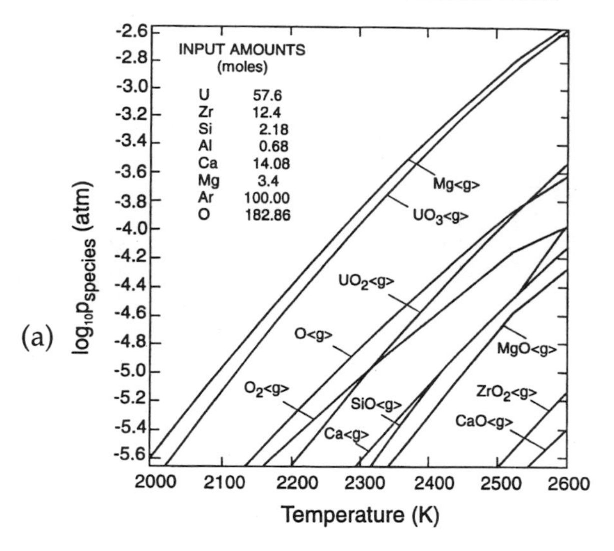

Corium-20 mol. % concrete: Non - Ideal model

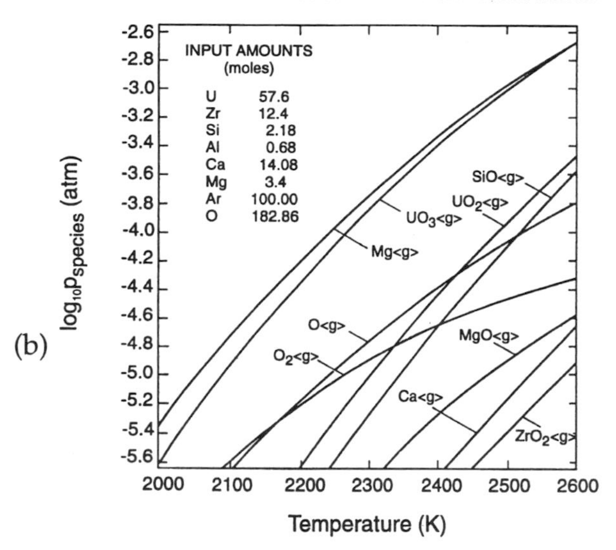

Fig. 15.4 The variation of the partial pressures of the dominant gas phase species with temperature for 30% oxidation/limestone concrete systems: (a) corium–20 mol% concrete: Ideal model; (b) corium–20 mol% concrete: Non-ideal model; (c) corium–80 mol% concrete: Ideal model; (d) corium-80 mol% concrete: Non-ideal model.

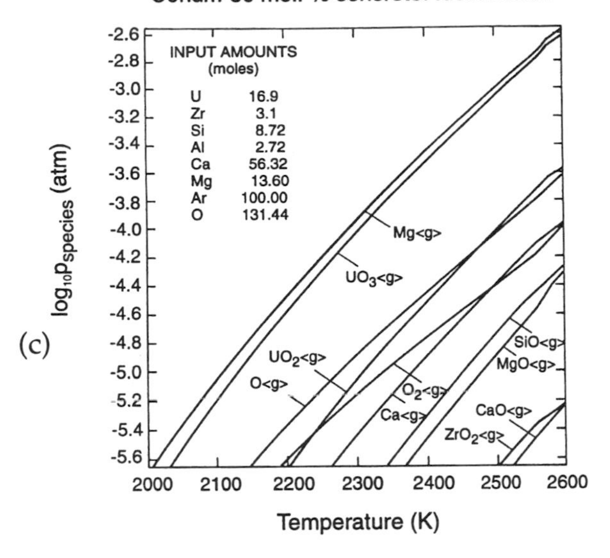

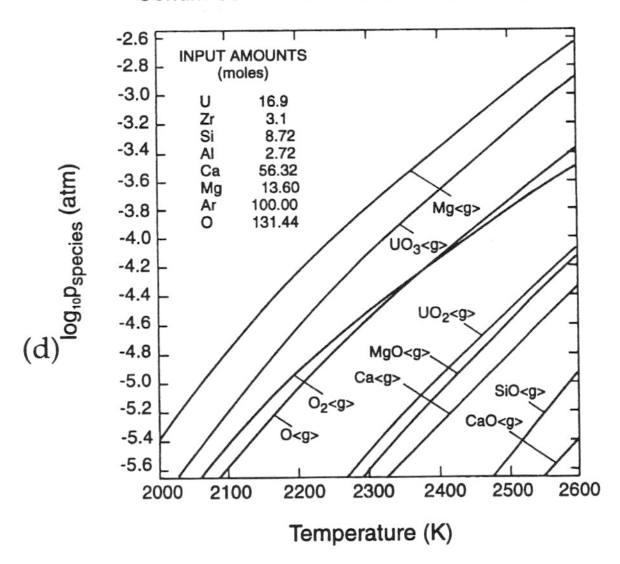

Fig. 15.4 (cont.)

solutions [84Col] and a database containing data for a simplified Gibbs energy model in the database are treated as ideal, both simple and complex constituents are included. Using the same test conditions and input amounts, two sets of calculations have been performed to compare the results of calculations carried out using the different approaches. One used the complex oxide solution model for the system UO_2–ZrO_2–SiO_2–CaO–MgO–Al_2O_3–BaO–SrO–La_2O_3, and the other used the ideal associated solution database.

Table 15.3 Comparison of release data from calculations using the full oxide model and the simplified model.

Species	Amount in the gas phase (moles)		Ratio of amounts
	M+	S+	M/S
Ba	1.900E-03	1.744E-03	1.1
Sr	2.518E-03	1.066E-04	24
U	2.252E+01	1.395E+01	1.6
Si	7.428E-03	3.061E-03	2.4
CO	4.007E+01	4.555E+01	-
CO_2	8.493E+01	7.945E+01	-
H_2	3.343E+00	2.946E+00	-
H_2O	4.790E+01	4.861E+01	-

+M = Full model S = Simplified model
Conditions for both calculations: Temperature = 2700K,

Input amounts (moles)

UO_2	783	Al_2O_3	3
ZrO_2	481	BaO	10
CaO	502	SrO	10
SiO_2	74	CO_2	125
MgO	15	H_2O	56

The releases for Ba, Sr, U and Si, using the two methods, are compared in Table 15.3. The amounts in the gas phase for all of these species are greater for the calculations using the oxide solution model, though the difference issignificant only for Sr. The values for the concrete decomposition gases are similar for the two calculations. The comparison of the amounts of the species in the gas phase is specific to the conditions of temperature and composition selected for these calculations. A sensitivity study is needed to determine the extent of the discrepancies for the range of conditions of the melt for MCCIs.

Conclusions

The erosion of the concrete basemat of a PWR due to the molten core–concrete interaction is a potential threat to containment integrity. In addition, fission products, fuel, and the components of the core structural materials and concrete would be released from the melt into the containment atmosphere by vaporisation. To model the thermophysical and chemical behaviour of the melt and hence the consequences of the interaction, it is necessary to have a good understanding of the phase equilibria in the gas phase and multi-component oxidic and metallic solution phases. Models are being developed, based on the assessment and optimisation of the available thermodynamic and phase diagram data, that can provide a detailed description of the composition of the melt for different conditions of temperature, corium and type of concrete.

The development of the thermodynamic database for the UO_2–ZrO_2–SiO_2–CaO–MgO–Al_2O_3–SrO–BaO–La_2O_3 system, based on the data for the binary and ternary sub-systems, is now complete. Part of the data for the solution model were obtained from a collaborative programme between the UK (involving AEA Technology and the National Physical Laboratory) and France (involving the CEA and Thermodata). Additional metal–oxide systems have also been studied to provide a model for the interaction between the immiscible oxide and metal solutions of the MCCI melt during the early phase of the interaction. At present, the systems U–UO_3, Zr–ZrO_2 and Si–SiO_2 have been evaluated. The data are consistent with the oxide solution models and the SGTE metal solution database.

A number of calculations of the phase constitution of the system as a function of composition and temperature have been performed. Reasonable agreement between the calculated and experimental data for the phase diagrams of the sub-systems was achieved. The oxide solution model has also been used to estimate the solidus–liquidus surfaces for oxide compositions of the full system that describe different extents of oxidation of the zirconium in the core debris and various concrete types. In addition to providing these temperatures, the model has been used to follow the evolution of the different phases in the system with temperature. In particular, the relative amounts and compositions of the liquid and solid phases in the melt have been calculated during the progression of the MCCI.

The oxide solution database has also been extended to include the gas phase species of the ten element system and MTDATA calculations have been performed to determine the extent of release of the components of the system by vaporisation. The results from the calculations have been compared with those from models that assume the oxide solutions are treated as ideal. A similar comparison has also been made between the non-ideal model approach and the method adopted to estimate the release of fission products during the progression of MCCI using a simplified ideal solution model. In the latter case, the amounts in the gas phase for Ba, Sr and U are greater for the MTDATA calculations using the oxide solution model; although the difference is significant only for Sr. The comparison of the amounts of the species in the gas phase is specific to the conditions of temperature and composition selected for these calculations. A sensitivity study is needed to determine the extent of the discrepancies between the non-ideal and simplified ideal solution models for the range of conditions of the melt for MCCIs.

Acknowledgements

Some of this work has been carried out by AEA Technology under contract for the Health and Safety Executive, UK, and forms part of a programme on nuclear safety research. Work at the NPL has been carried out as part of a multi-client programme organised by the Mineral Industry Research Organisation with support from the UK Department of Trade and Industry. The collaboration of Thermodata in the development of the data is gratefully acknowledged.

References

77Luk	H.L.Lukas, E.T.Henig and B.Zimmerman: *Calphad* **1**, 1977, 225.
82Som	F.Sommer: *Z. Metallkde.* **73**, 1982, 72
84Col	R.K.Cole, D.P.Kelley and M.A.Ellis: 'CORCON-Mod 2: A Computer Program for Analysis of Molten Core-Concrete Interactions', NUREG/CR-3920, August 1984.
85Hil	M.Hillert, B.Jansson, B.Sundman and J.Ågren: *Metall. Trans.* **16A**, 1985, 261.
85JAN	JANAF Thermochemical Tables, Third Edition, *J Phys. Chem. Ref. Data* **14**, 1985 Supplement No. 1.
86But	A.T.D.Butland, M.A.Mignanelli, P.E.Potter,P.N.Smith: 'The vaporization of chemical species and the production of aerosols during a core debris-concrete interaction', *Proc. CSNI Specialist Meeting on Core Debris-Concrete Interactions,* Palo Alto, USA, 1986.
86Smi	P.N.Smith, A.T.D.Butland, P.E.Potter, M.A.Mignanelli,M.A. and G.J.Roberts: 'The importance of core/concrete aerosol production and some containment heat sources to the source term', *Proc. Symp. on Source Term Evaluation for Accident Conditions,* Columbus, Ohio, 1985, IAEA-SM-281/32, 1986.
87Ans	I.Ansara, and B.Sundman in: *Computer Handling and Dissemination of Data,* ed. P.S. Glaeser, CODATA 1987.
87Bha	A.S.Bhansali and A.K.Mallik.: *Calphad* **11**, 1987, 105.
90Cor	E.H.P.Cordfunke, R.J.M.Konings, G.Prins, P.E.Potter and M.H.Rand: *Thermochemical Data for Reactor Materials and Fission Products,* Elsevier Science Publishers, Amsterdam, 1990.
92Che	P.Y.Chevalier: *J. Nucl. Mats.* **186**, 1992, 212.
92Hor	P.J.Horrocks: AEA Technology, Private Communication, 1992.
92Luk	H.L.Lukas: MPI, Stuttgart, Private Communication, 1992.
92Mig	M.A.Mignanelli: 'A study of the parameters influencing the release of species by vaporisation during core-concrete interactions', *Proc. of Second OECD (NEA) CSNI Specialist Meeting on Molten Core Debris–Concrete Interactions,* Karlsruhe, Germany, 1-3 April, 1992, 287, NEA/CSNI/R(92)10, 1992.
93Bal	R.G.J.Ball, M.A.Mignanelli T.L.Barry and J.A.Gisby: *J. Nucl. Mats.* **201**, 1993, 238.
93Ran	M.H.Rand: AEA Technology, Private Communication, Mar 1993.
93Roc	M.F.Roche, L.Leibowitz, J.K.Fink, and L.Baker Jr: NUREG/CR-6032, June 1993.

16 Pyrometallurgy of Copper-Nickel-Iron Sulphide Ores: The Calculation of Distribution of Components between Matte, Slag, Alloy and Gas Phases

TOM I. BARRY*, ALAN T. DINSDALE¶, SUSAN M. HODSON¶, JEFF R. TAYLOR‡

Introduction

A basic principle of many processes for extracting copper and nickel from sulphide ores is to overlay the molten ore or matte with a lime-silica-iron oxide slag and blow with air or oxygen. Some of the iron present in the ore oxidises and, as a result, partitions to the slag phase, while some of the sulphur is carried off as sulphur dioxide. In this example of the application of computer assisted thermochemistry a series of calculations is described which model the phase equilibria and partition of elements between the matte, slag, alloy and gas phases at various stages of the pyrometallurgical process. The results agree well with practical experience and also provide an insight into the detailed phase relationships to be expected during such processes. The subsequent processing depends very much on the composition of the ore, including the amounts of impurities, some of which may themselves be of value.

The following discussion does not relate to any particular process and is intended only to illustrate the calculation procedures. Not all factors have been taken into account in the thermodynamic modelling. For example, the solution of Cu, Ni and S in the slag is not considered. Three types of processes are considered: these are more fully described by Dinsdale, Hodson, Barry and Taylor [88Din2].

- Partition of iron to the slag phase and sulphur to the gas phase while blowing the matte.
- The effect of sulphur content on phase separation in the liquid mattes especially those with low nickel content.
- Crystal transformation during solidification of a matte containing reduced amounts of sulphur and iron, which can be used to effect a further partition of the components.

* Amethyst Systems, 2 Marlingdene Close, hampton, Middlesex TW12 3BJ, UK
¶ Centre for Materials Measurement and Technology, National Physical Laboratory, Middlesex TW11 0LW, UK
‡ Johnson Matthey Technology Centre, Blount's Court, Sonning Common, Reading, Berkshire RG4 9NH

Table 16.1 Initial Compositions of the Matte and Slag phases.

	matte A		matte B			slag	
Component	wt%	moles	wt%	moles	Component	wt%	moles
Ni	26.7	1.0	7.9	0.3	CaO	10.0	1.18
Fe	25.4	1.0	25.0	1.0	'FeO'	60.0	5.52
Cu	28.9	1.0	48.4	1.7	SiO_2	30.0	3.33
S	18.9	1.3	18.6	1.3			
Total mass /g	219.7		223.0			662.9	

Blowing the matte

For these calculations data are required for the matte, slag and gas phases. The plant operator needs to process ores of a range of compositions and to achieve the desired result he can vary the temperature, the composition of the slag and the amount of air/oxygen used in the blow. Using these variables he can achieve some degree of independent control over the amounts of iron and sulphur left in the matte. The partial pressure of sulphur dioxide in the off gas can be used as a monitor of the process.

The initial compositions of the mattes and slags used in the calculations below are given in Table 16.1. The mass of the slag is made three times that of the matte so that its composition does not change markedly. In practice the slag would be replenished either from time to time or continuously. In continuous processing the slag could be made to flow countercurrent to the matte. Two different matte compositions are used to illustrate different aspects of processing.

Figure 16.1 shows the calculated effect of blowing a matte (matte A) which originally contained equi-molar proportions of nickel, iron and copper. Subsequently the oxygen potential and partial pressure of sulphur dioxide rise and the amount of sulphur remaining in the matte falls. In agreement with practical experience, calculations show that initially the added oxygen is consumed mainly in the process of oxidation of iron, which is transferred to the slag. As a result the proportion of sulphur in the matte actually increases. In Figure 16.1 oxygen is added progressively until a substantial pressure of sulphur dioxide develops. Most of the iron is removed from the matte during this phase.

In the second phase of the process the partial pressure of sulphur dioxide is considered to be maintained at 0.1 atm. Figure 16.2 shows the calculated relationship between the amounts of iron and sulphur in the matte under these chosen conditions. The relationship between these two variables is a function of the composition of the slag and the relative proportions of nickel and copper in the matte, the temperature and the partial pressure of sulphur dioxide. A knowledge of this relationship is important to the control of the blowing process in order to prepare the matte for subsequent processing.

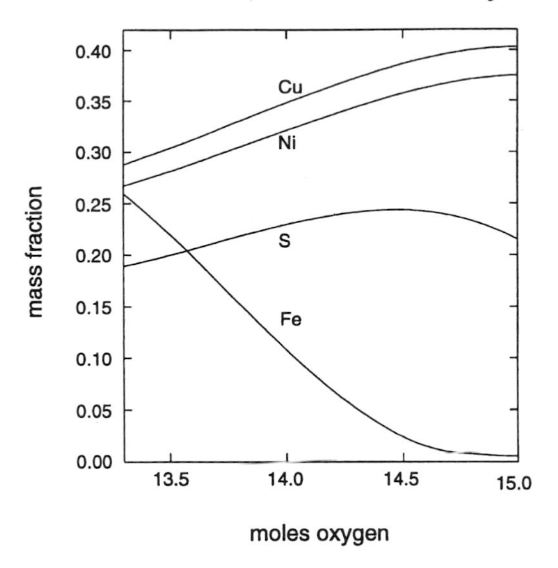

Fig. 16.1 Composition of matte A as a function of oxygen content of the system at 1573 K. The starting amount of oxygen before blowing is 13.36 mol.

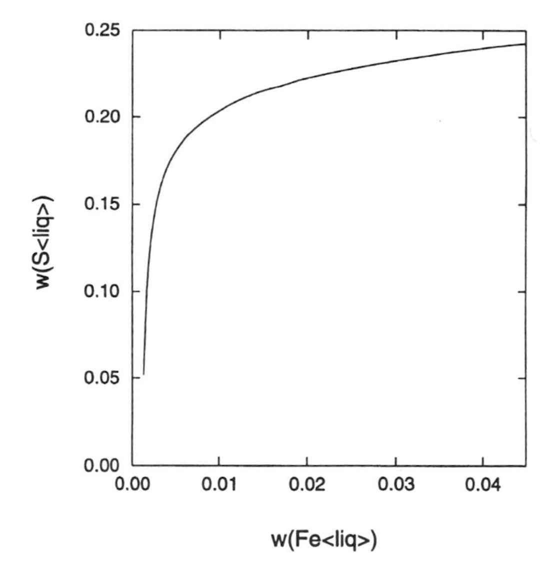

Fig. 16.2 Interdependence of the weight fraction of iron and sulphur in the liquid for a particular nickel-copper ratio, slag composition and sulphur dioxide pressure.

Good agreement between calculated and experimental results has been demonstrated elsewhere [88Din1, 90Tay1]. The strongly non-ideal behaviour is reflected in both the calculated and experimental results.

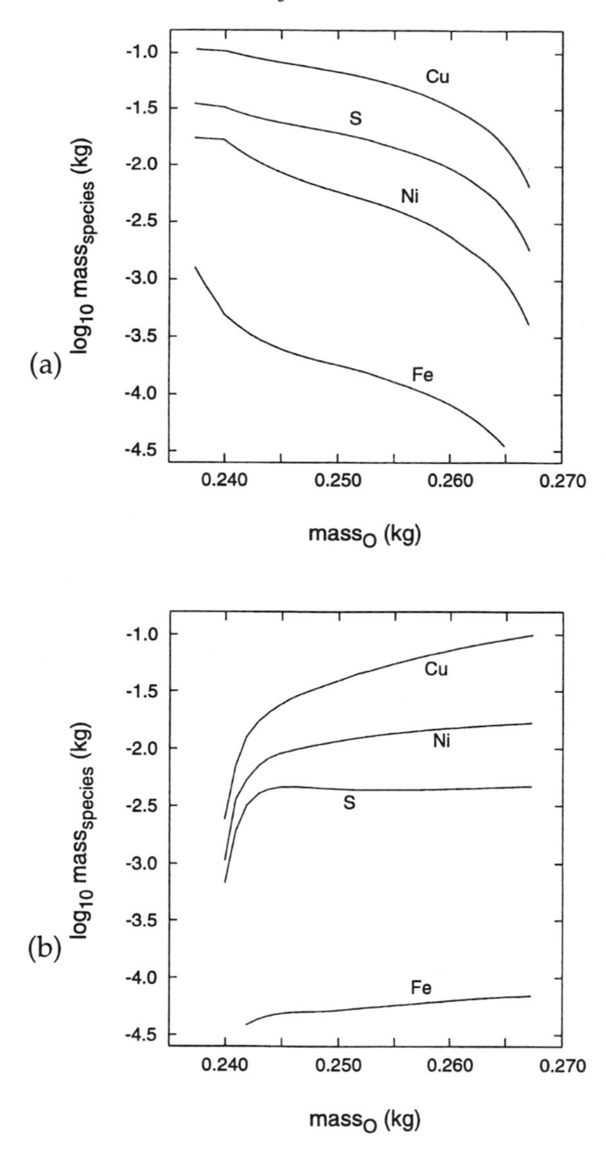

Fig. 16.3 Two-liquid formation in matte B as a function of oxygen content of the system at 1573 K. Compositions of (a) the sulphur-rich and (b) the sulphur-poor liquid.

Phase separation in the matte

There is a considerable tendency for mattes, particularly those of low nickel content, to unmix into sulphur-rich and sulphur poor liquids. This is exploited in the Noranda process in which copper rich mattes are blown down to a sulphur level at which a sulphur poor liquid enriched in nickel separates.

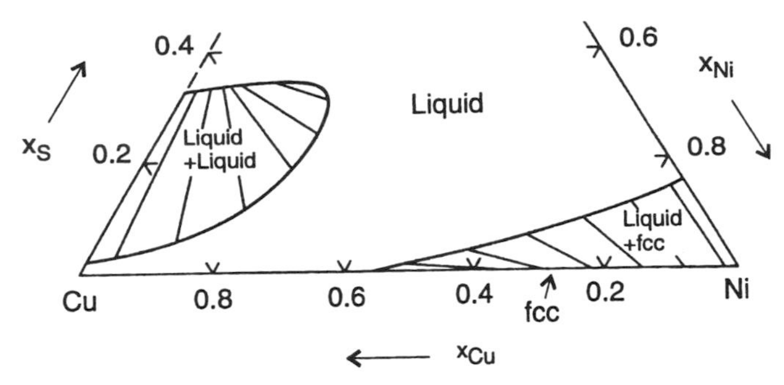

Fig. 16.4 Calculated phase diagram for the Cu-Ni-S system for 1573 K.

A feature of copper-rich, nickel-poor mattes is that they tend to develop liquid–liquid immiscibility. Thus at 1573 K matte B exists as two liquids. As the amount of iron is reduced, the two liquids become miscible but immiscibility again develops as the amount of sulphur declines and a sulphur-poor phase separates.

The proportions of the two phases and the distribution of components between them in this second regime are illustrated in Figure 16.3. The proportion of iron in the matte has been reduced to low levels and thus the composition lies nearly within the Cu-Ni-S ternary system of Figure 16.4. In this diagram, the position of the miscibility gap, which is not perfectly in accord with the experimental data reviewed by Chang, Neumann and Choudray [79Cha], suggests that for the compositions of Figure 16.3 phase separation should begin at about 30 wt% sulphur. Because the tielines in Figure 16.4 run at an angle to lines of constant copper-nickel ratio, enrichment of nickel occurs in the sulphur-poor phase. If a differential flow is established between the phases it is possible to reduce the level of nickel in the sulphur-rich phase (which will follow the upper edge of the miscibility gap) to low levels. This possibility is exploited in the Noranda process, the thermodynamics of which has been explored by Nagamori and Mackey [78Nag].

It is instructive to examine the results of a single step in the calculations displayed graphically in Figure 16.3. This is given in Table 16.2.

Solidification and recrystallisation

Further separation of the components occurs during crystallisation of the matte on cooling. It may sometimes be advantageous not to reduce the iron to very low levels; among other things this avoids loss of values to the slag. Figure 16.5a shows the predicted sequence of phases formed during cooling of a matte of original composition A in which the amount of iron has been reduced to 10 wt% and the sulphur to 22.3 wt%. The compositions of the fcc phase are given in Figure 16.5b and the distribution of components between the phases is given in Figure 16.6.

Table 16.2 Edited results showing the composition of the slag and liquid phases for a single set of conditions. Ferric iron in the slag is identified as $Fe_{2/3}O$.

Temperature = 1573.00 K
Fixed gas pressure = 1.013250E+05 Pa Calculated gas volume = 7.172848E-01 m3

	Phase	Species	Amount mole	Mole fraction
Phase LIQUID				
Ni(LIQ)	4	1	0.19876	0.20751
Fe(LIQ)	4	2	0.00095	0.00100
Cu(LIQ)	4	3	0.61835	0.64556
S(LIQ)	4	4	0.13979	0.14594
Phase total is			0.95786	1.00000
Phase LIQUID				
Ni(LIQ)	4	1	0.10124	0.05658
Fe(LIQ)	4	2	0.00323	0.00181
Cu(LIQ)	4	3	1.08153	0.60446
S(LIQ)	4	4	0.60326	0.33716
Phase total is			1.78926	1.00000
Phase SLAG				
$Si_{0.5}O$	11	1	1.44802	0.16179
$Fe_{2/3}O$	11	2	0.04811	0.00538
CaO	11	3	0.00017	0.00002
FeO	11	4	1.90744	0.21312
$Si_{0.5}OFe_{2/3}O$	11	5	0.20923	0.02338
$Si_{0.5}OCaO$	11	6	1.10093	0.12301
$Si_{0.5}OFeO$	11	7	3.84182	0.42925
$Fe_{2/3}OCaO$	11	8	0.02883	0.00322
$Fe_{2/3}OFeO$	11	9	0.31542	0.03524
CaOFeO	11	10	0.05007	0.00559
Phase total is			8.95004	1.00000
Phase GAS				
Cu<g>	12	1	0.00012	0.00002
N_2<g>	12	4	4.99999	0.89976
OS<g>	12	8	0.00114	0.00021
O_2S<g>	12	10	0.55571	0.10000
S_2<g>	12	12	0.00003	0.00001
Phase total is			5.55703	1.00000

Component	Chem. Pot.	Activity		Moles
Ni	-1.023565E+05	3.991462E-		3.000000E-01
Fe	-1.566855E+05	6.267199E-		6.520000E+00
Cu	-9.278681E+04	8.296679E-		1.700000E+00
S	-2.181046E+05	5.722415E-		1.300000E+00
O	-2.814964E+05	4.493439E-		1.560896E+01
Si	-5.128231E+05	9.356250E-		3.300000E+00
Ca	-5.594041E+05	2.656610E-		1.180000E+00
N	-1.716695E+05	1.993016E-		1.000000E+01

Gibbs Energy = -9.956548E+06 J

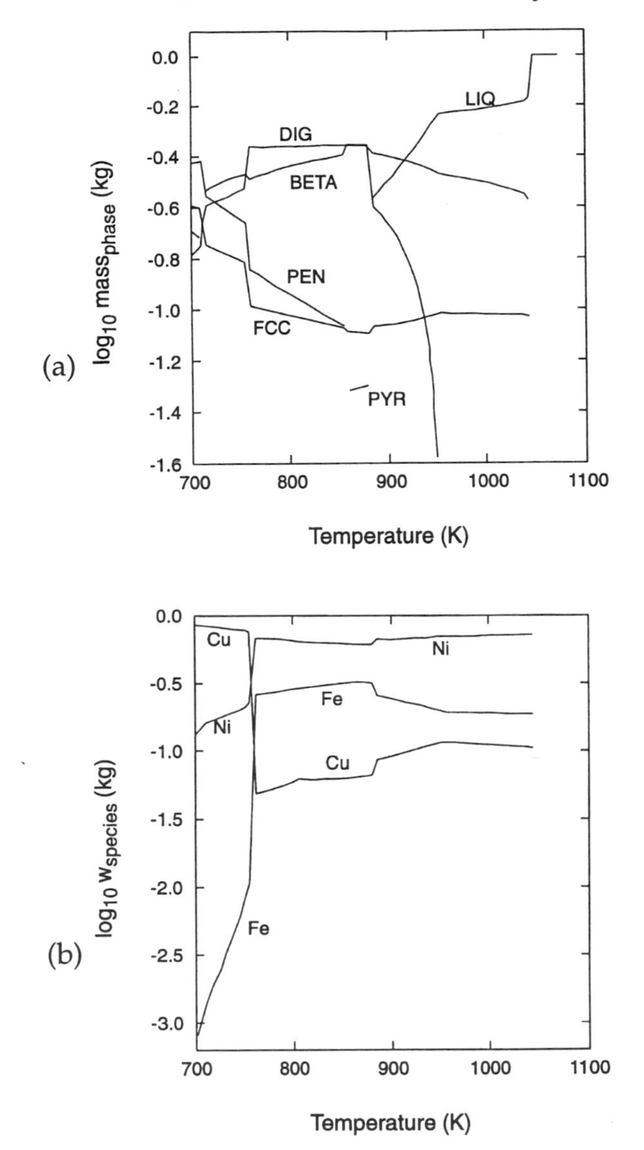

Fig. 16.5 Phase equilibria as a function of temperature in matte A depleted in sulphur and iron. The composition is Ni 0.325, Fe 0.1, Cu 0.352, S 0.223 parts by weight. (a) mass of phases, cf Table 16.3, (b) composition of the fcc phase.

Solidification sets in at 1050 K with the formation of the beta and metallic fcc phases. Nickel is strongly segregated to the beta phase, which approximates to Ni_3S_2 in composition. Plots similar to Figure 16.5b show that the iron maintains its concentration in the diminishing amount of liquid, whilst the concentration of copper is enhanced. At 950 K digenite precipitates, depleting the liquid in copper and sulphur and leading to the disappearance of the liquid at 900 K. At this temperature iron is partitioned to the beta and transient pyrrhotite phase, which has only a small temperature range of stability before the emergence of pentlandite. Not obvious in these plots but evident in the copper-nickel phase diagram is a tendency to immiscibility in the fcc phase at low temperature. This contributes to a further transformation at 750 K, in which the amount of copper combined with sulphur as digenite sharply declines, whereas that of nickel and iron combined with sulphur in pentlandite increases.

The calculations relate only to a single composition. Nevertheless they demonstrate that the crystallisation processes in a multiphase matte can be very complex and that calculations based on thermodynamic data can contribute greatly to the selection and optimisation of processing methods.

Thermodynamic models and data

(The phases and thermodynamic models are given in Table 16.3.)

The data for the sulphide phases were derived exclusively from the assessments of Dinsdale [82Din, 84Din] and Fernandez Guillermet and colleagues [81Fer]. The metal/matte liquid is described by a two-sublattice model with variable site fractions similar to that of Hillert and colleagues [85Hil]. Both binary and ternary experimental data were taken into consideration in the assessment. Most of the several crystalline phases are modelled as solid solutions. The data for the slag phase used for this work are based with some modification on the work of Gaye and Welfringer [84Gay] and Taylor and Dinsdale [90Tay], who used the cellular model of Kapoor and Frohberg [71Kap]. Other models used elsewhere for slags include the sublattice model, the quasichemical model of Pelton and Blander [86Pel] and the associated solution model [93Bar]. Data for the gaseous species were obtained from the SGTE pure substance database [85Bar, 87Ans]. All the data were converted to the G-Hser formalism, in which the Gibbs energy

Table 16.3 Phases and Thermodynamic Models.

gas	Ni–Fe–Cu–O–S–N	ideal gas
slag	$CaO–FeO–Fe_2O_3–SiO_2$	cellular model
liquid	Ni–Fe–Cu–S	variable sublattice
fcc	Ni–Fe–Cu	Redlich Kister
bcc	Ni–Fe–Cu	Redlich Kister
beta('Ni3S2')	Ni–Fe–Cu–S	Redlich Kister
pyrrhotite	(Ni,Fe)S	Redlich Kister
pentlandite	$(Ni,Fe)S_{0.889}$	Redlich Kister
digenite	Cu_2S	Stoichiometric
heazlewoodite	Ni_3S_2	Stoichiometric
millerite	NiS	Stoichiometric
Ni7S6	Ni_7S_6	Stoichiometric

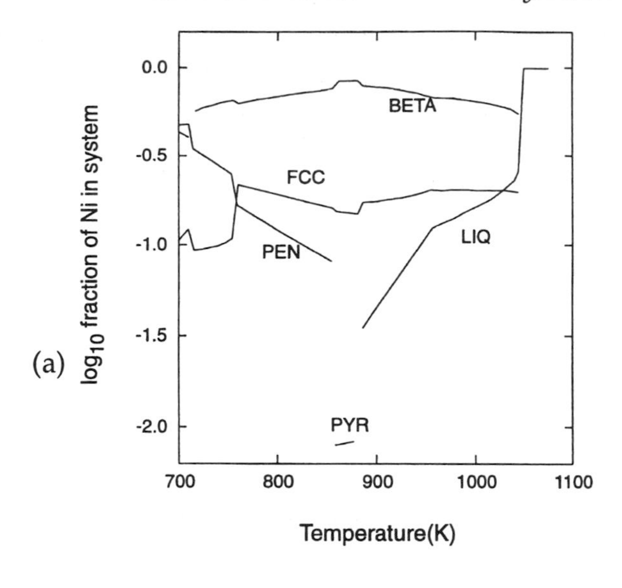

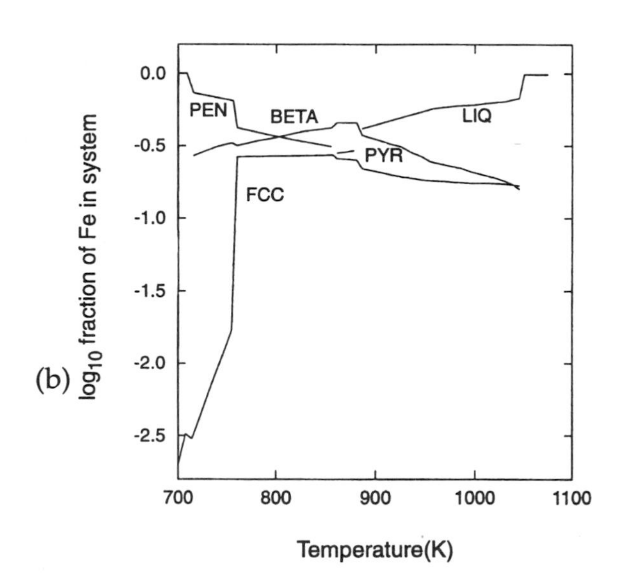

Fig. 16.6 Distribution of the components of the matte between the phases as listed in Table 16.3. The total mass of each phase is given in Figure 16.5a.

data for all substances are referred to the properties of the pure elements in their stable states at 298.15 K and 101325 Pa, rather than to the elements in a specified phase at current temperature.

Acknowledgements
The authors gratefully acknowledge the help of T. G. Chart for his advice and help in the critical assessment of data and to R. H. Davies and J. A. Gisby for

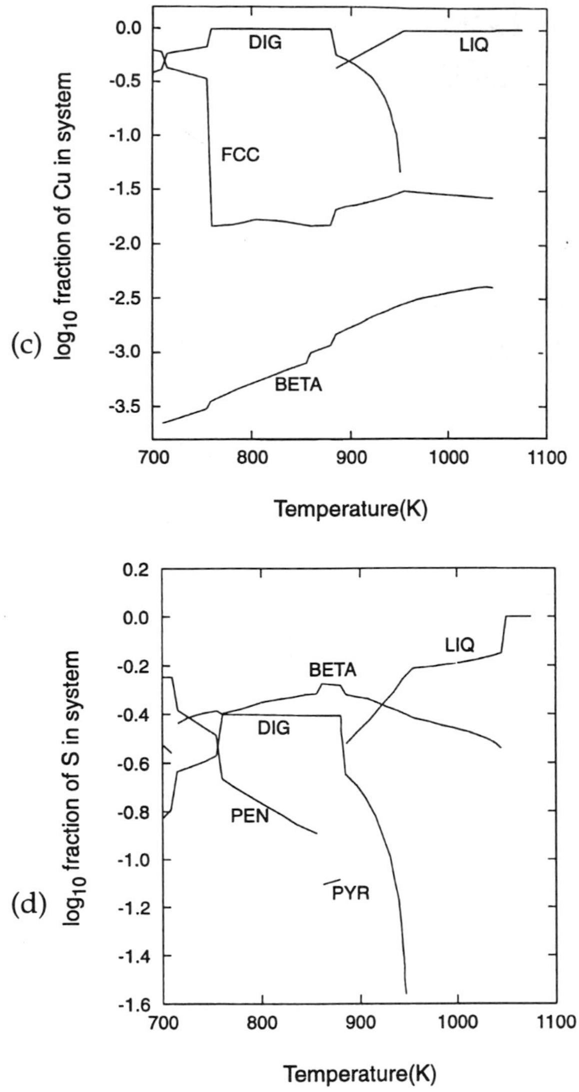

Fig. 16.6 cont.

assistance with the calculation of the phase equilibria. This work was sponsored in part by Matthey Rustenberg Refiners. The calculations for this case study have been performed using MTDATA.

References

71Kap M.L.Kapoor and G.M.Frohberg: 'Theoretical treatment of activities in silicate melts', *Proceedings Symp. Chemical Metallurgy of Iron and Steel*, Sheffield, 1971, 17-22, The Iron and Steel Institute, London.

78Nag N.A.Nagamori and P.J.Mackey: 'Thermodynamics of copper matte converting, Part 1: Fundamentals of the Noranda Process', *Metall. Trans.* **9B**, 1978, 255–265.

79Cha Y.A.Chang, J.P.Neumann and U.V.Choudray.: 'Phase Diagrams and Thermodynamic Properties of Ternary Copper–Sulphur–Metal Systems', INCRA Monograph VII, *The Metallurgy of Copper,* INCRA, New York, 1979.

81Fer A.Fernandez Guillermet, M.Hillert, B.Jansson and B.Sundman: 'An assessment of the Fe-S system using a two-sublattice model for the liquid phase', *Metall. Trans.* **12B**, 1981, 745–754.

82Din A.T.Dinsdale, T.G.Chart, T.I.Barry and J.R.Taylor: 'Phase equilibria and thermodynamic data for the Cu-S system', *High Temperatures – High Pressures* **14**, 1982, 633–640.

84Din A.T.Dinsdale: 'The generation and application of thermodynamic data', Thesis, Brunel University, 1984.

84Gay H.Gaye and J.Welfringer.: 'Modelling of the thermodynamic properties of complex metallurgical slags', in H.A.Fine,H.A. & D.R.Gaskell (Eds), *Proceedings 2nd Internat. Symp., Metal Slags and Fluxes,* Lake Tahoe, Metall. Soc. AIME, New York, 1984, 357–375.

85Bar T.I.Barry: 'High temperature inorganic chemistry and metallurgy', in T.I.Barry (Ed), *Chemical Thermodynamics in Industry: Models and Computation,* Chap. 1, 1–39. Blackwell Scientific Publications, Oxford, 1985.

85Hil M.Hillert, B.Jansson, B.Sundman and J.Ågren: 'A two-sublattice model for molten solutions with different tendency to ionization', *Metall. Trans.* **16A**, 1985, 261–266.

86Pel A.D.Pelton and M.Blander: 'Thermodynamic analysis of ordered liquid solutions by a modified quasichemical approach - Application to silicate slags', *Metall. Trans.* **17B**, 1986, 805-815.

87Ans I.Ansara and B.Sundman: 'The Scientific Group Thermodata Europe', in P.S.Glaeser (Ed), *Computer handling and dissemination of data,* 154–158, Elsevier Science Publishers, Amsterdam, 1987.

88Din1 A.T.Dinsdale, S.M.Hodson and J.R.Taylor: 'Application of MTDATA to the modelling of slag, matte, metal, gas phase equilibria', *Proceedings 3rd Internat. Conf. Molten slags and fluxes,* Strathclyde, Institute of Metals, London, 1988.

88Din2 A.T.Dinsdale, S.M.Hodson, T.I.Barry, and J.R.Taylor: 'Computations using MTDATA of metal–matte–slag–gas equilibria', *Proceedings 27th Annual Conference of Metallurgists,* CIM Montreal, 1988.

90Tay1 J.R.Taylor and A.T.Dinsdale: 'Application of the calculation of phase equilibria to the pyrometallurgical extraction from sulphide ores', *Proc. Conf. User Aspects of Phase Diagrams,* 25–27 June

1990, Petten, Institute of Metals.

90Tay2 J.R.Taylor and A.T.Dinsdale: 'Thermodynamic and phase diagram data for the system CaO-SiO_2 system', *Calphad* **14(1)**, 1990, 71–88.

93Bar T.I.Barry, A.T.Dinsdale, and J.A.Gisby: "Predictive Thermochemistry and Phase Equilibria of Slags", JOM, April 1993, 32-38.

17 High-Temperature Corrosion of SiC in Hydrogen–Oxygen Environments

KLAUS G. NICKEL*, HANS L. LUKAS* AND GÜNTER PETZOW*

Abstract

A thermodynamic analysis of parts of the Si–C–H and Si–C–O–H systems reveals main gas species for the hot-gas corrosion of SiC. The analysis is combined with simple kinetic models to assess the corrosion kinetics at high temperatures in environments with static, flowing and extremely fast flowing atmospheres.

Analysis and models allow to assess corrosion processes chemically (main reaction determination) and kinetically (extrapolation of corrosion kinetics at steady state or moderate flow conditions to temperatures of interest). In conditions of extremely fast gas flows safe temperature limits can be estimated. In addition we predict the corrosion style (grain boundary destruction, bulk corrosion).

Introduction

SiC is used commercially for high temperature applications like heating elements because of its resistance to oxidation corrosion to at least 1600°C. It is also one of the very few materials which have the capability to retain strength at high temperatures. Thus it is a prime candidate for currently developed applications such as heat engine linings or heat exchangers.

In such applications gas temperatures may exceed 3000°C and imply extremely fast gas flow (e.g. in turbines of super-fast planes). The prevailing atmosphere in such engines depends on the fuel used. Hydrogen-oxygen is known as a very efficient fuel system in rockets. Heat exchangers may also be exposed to various atmosphere compositions including steam at high temperatures. The resistivity of SiC against hydrogen-oxygen corrosion is thus a limiting factor for the applicability of this material.

We have simulated three basic types of gaseous corrosion situations by computing models for static, moderate and extremely high flow rates of the atmosphere. The importance of the analysis of the application environment is demonstrated. The validity of the use of simple models is evaluated by comparing the predicted damage to those of experimental data.

* Pulvermetallurgische Laboratorium, Institut für Wekstoffwissenschaft, Max-Planck-Institut für Metallforschung, Heisenbergstraße 5, D–70569, Germany.

It is obvious from the modelling procedure that the prediction of the corrosion kinetics is based on thermodynamic analysis, in particular the calculation of partial pressure of gas species. Hence we present the main points of a thermodynamic analysis of parts of the systems Si-C-H and Si-C-O-H, which are of interest for the problem of steam, 'dry' and 'wet' hydrogen corrosion of SiC.

The partial pressures over SiC + C are a model system for pressureless sintered SiC ('SSiC'), which contains usually some free carbon. Other types of commercially free available SiC ('SiSiC') contain free silicon. Hence partial pressures over Si + SiC are relevant for this type of material.

Thermodynamically equilibria which are not invariant at constant pressure and temperature can be calculated but are of little general significance because of their compositional dependence. However, some bulk compositions are of technical interest, because they refer to an existing material. Such a case is pure SiC, which may be purchased as SiC from chemical vapour deposition production or will be produced in a dynamic process such as the removal of carbon from SSiC. Hence we have included some calculations for this type of material.

Models for the corrosion process

Corrosion processes which are controlled by the emission of gaseous particles from a material are strongly dependent on the boundary conditions of the system. We may distinguish three basic types of corrosion situations: static, flowing and extremely fast flowing atmosphere.

A static atmosphere with a defined available gas volume has a defined maximum loss of material to the atmosphere, because at equilibrium the equation of state for gases

$$n = \frac{PV}{RT} \tag{17.1}$$

defines the number of moles of the individual species, which may enter the available gas chamber volume. Because this type of equilibrium is rarely overstepped, the value is a maximum, which will however be in good agreement with reality at high temperatures.

We may thus calculate the amount of any element in the atmosphere from the thermodynamic analysis of this relation by adding the contribution of each gas species. From here we may calculate further the amount of elements from SiC now present in gas form.

While this procedure may be performed by a computer, in practice this is hardly necessary, because the gas phase is usually strongly dominated by a single species at a given temperature; the sum of all minor species will thus only add up to a minor or negligible fraction of the total amount of species containing elements of the material to be corroded.

From the dominating gas species we can immediately deduce the main reaction: If e.g. the only condensed species is SiC and the main gas species is mono-atomic Si it follows that SiC→ Si↑ + C is the dominating reaction. A list of important reactions in the Si–C–H(–O) systems is given in Table 17.1.

Knowing the dominating reaction we have also defined the amount Δm of SiC destroyed, because the reaction equation relates the moles of evaporating and destroyed species:

$$\Delta m = M \frac{n_{gas}}{u} \tag{17.2}$$

where M is the molecular weight of SiC and u the factor relating the moles of evaporating species and destroyed SiC in the reaction equation. We thus have

$$\Delta m = \frac{MP_iV}{uRT} \tag{17.3}$$

or, if we know the exposed surface a, the corrosion depth via the density relation:

$$X_l = \frac{MP_iV}{a\rho uRT} \tag{17.4}$$

The corrosion depth in equation (17.4) is mainly dependent on V and P_i and cannot be large unless the partial pressure or the volume of the gas chamber is huge. The introduction of a factor u implies that the corrosion depth is defined as the depth to which the material is destroyed and not by the width of shape change, because the corrosion may leave a (non-protective) layer of residue e.g. of carbon, if the main reaction is reaction (R7) (Table 17.1).

A corrosion rate (corrosion depth per time unit) follows from equation (17.4) directly, if we assume a regular exchange of the atmosphere (i.e. a regular removal of the species containing Si or C) with time. We get the total amount of corrosion by multiplying with the exchange rate ν (the number of exchanges per time unit):

$$\mathring{X}_l = \frac{MP_iV}{a\rho uRT}\nu \tag{17.5}$$

Since a flow rate of an atmosphere is defined by

$$\mathring{V} = V\nu \tag{17.6}$$

we also have

$$\overset{\circ}{X}_l = \frac{MP_i\overset{\circ}{V}}{a\rho uRT} \tag{17.7}$$

Table 17.1 Reaction equations.

Si	$\rightleftharpoons$	$Si\uparrow$	(R1)
$\frac{1}{2}H_2 + Si$	$\rightleftharpoons$	$SiH\uparrow$	(R2)
$2\,H_2 + Si$	$\rightleftharpoons$	$SiH_4\uparrow$	(R3)
$\frac{1}{2}H + 2\,C$	$\rightleftharpoons$	$C_2H\uparrow$	(R4)
$H_2 + 2\,C$	$\rightleftharpoons$	C_2H_2	(R5)
$2\,H_2 + C$	$\rightleftharpoons$	$CH_4\uparrow$	(R6)
SiC	$\rightleftharpoons$	$Si\uparrow + C$	(R7)
$2\,SiC$	$\rightleftharpoons$	$SiC_2\uparrow + Si$	(R8)
$2\,SiC$	$\rightleftharpoons$	$Si_2C\uparrow + C$	(R9)
$3\,SiC$	$\rightleftharpoons$	$SiC_2\uparrow + Si_2C\uparrow$	(R10)
$\frac{1}{2}H_2 + SiC$	$\rightleftharpoons$	$SiH\uparrow + C$	(R11)
$H_2 + 2\,SiC$	$\rightleftharpoons$	$C_2H_2\uparrow + 2\,Si$	(R12)
$H_2 + 2\,SiC$	$\rightleftharpoons$	$C_2H_2\uparrow + 2\,Si\uparrow$	(R13)
$2\,H_2 + SiC$	$\rightleftharpoons$	$CH_4\uparrow + Si$	(R14)
$2\,H_2 + SiC$	$\rightleftharpoons$	$SiH_4\uparrow + C$	(R15)
$H_2O + Si$	$\rightleftharpoons$	$SiO\uparrow + H_2$	(R16)
$H_2O + C$	$\rightleftharpoons$	$CO\uparrow + H_2$	(R17)
$2\,H_2O + Si$	$\rightleftharpoons$	$SiO_2\uparrow + 2\,H_2$	(R18)
$2\,H_2O + C$	$\rightleftharpoons$	$CO_2\uparrow + 2\,H_2$	(R19)
$2\,H_2O + SiC$	$\rightleftharpoons$	$SiO\uparrow + CO\uparrow + 2\,H_2$	(R20)
$H_2O + H_2 + SiC$	$\rightleftharpoons$	$CH_4\uparrow + SiO\uparrow$	(R21)

or, in terms of a mass flux J

$$J = \mathring{X}_l \, \nu = \frac{MP_i\mathring{V}}{a\rho uRT} \tag{17.8}$$

We have formulated the mass flux from equilibrium assumptions. Hence the calculation will give precise values if equilibrium is obtained and kinetic hindrances are negligible. If these conditions are not met we need to experimentally determine an effective flow rate , which allows to keep the form of equations (17.7) and (17.8) and give good corrosion predictions.

At extremely high gas flow rates the physics of the corrosion process are ill known and the assumption of obtained equilibrium is certainly not valid. As the most extreme case of gaseous decomposition we consider the removal of any gas species particle as soon as it is formed without being slowed down by any physical or chemical interaction. In this case the speed of removal is solely controlled by the speed of the gas particles, which is known from

$$\bar{\nu} = \sqrt{\frac{8kT}{\pi m}} \tag{17.9}$$

Since the molecular flux of gas is

$$\Phi = \frac{1}{4}\frac{N}{V}\nu \tag{17.10}$$

and equation (17.1) is still valid we may calculate the maximum possible flux as

$$J = \frac{P_i}{\sqrt{2\pi RTM}} \tag{17.11}$$

It should be emphasised that this is a worst-case of calculation, which will only be approached in extreme conditions such as ultra-high vacuum. We wanted to evaluate conditions of extreme cases, which are not easily accessible to experimental testing.

Thermodynamic Analysis

Thermodynamic analysis were performed employing the programs PMLFKT [82Luk] and SOLGASMIX [75Eri]. Using the SGTE data set [87Ans] as input parameters we considered the species listed in table 17.2.

Table 17.2 Species considered in the thermodynamic calculations.

gas species:	C, C^+, C^-, C_2^-, C_3, C_4, C_5, H, H^-, H^+, H_2, H_2^+, H_2^-, Si, Si^+, Si_2, Si_3, C_2H, CH, CH^+, C_2H_2, C_2H_4, C_2H_6, C_3H_8, CH_3, CH_4, SiC, SiC_2, Si_2C, SiH, SiH^+, SiH_4, Si_2H_6, O, O^+, O^-, O_2, O_2^+, O_2^-, O_3, C_2O, CO, CO_2, CO_2^-, C_3O_2, OH, OH^+, OH^-, HO_2, H_2O, H_3O^+, HCO, HCO^+, H_2CO, H_4CO, SiO, SiO_2
condensed species:	C (graphite), Si, Si (liquid), α-SiC, β-SiC,·SiO_2 (quartz), SiO_2 (tridymite), SiO_2 (cristoballite), SiO_2 (liquid)

In the corrosion models we refer to reaction equations. The relevant reaction equation is defined (see above) by those gas species, which have the highest partial pressures. Hence in plot of partial pressures of gas species vs. temperature we only need to show these species. The resulting plot is thus a 'maximum partial pressure surface'; in the following we abbreviate this type of plot 'MPPS'.

Si-C-H systems

The equilibria Si–C gas and Si–SiC gas are invariant at constant temperature and pressure. Therefore they can be calculated at constant total pressure for each temperature. All calculations in this work are for a total pressure of 1 bar.

Figures 17.1a and b show the MPPS for Si–SiC gas and SiC–C gas equilibria in the Si-C-H system. All stable two phase equilibria between SiC and gas lie between these three phase equilibria.

In the SiC-C-gas equilibria up to about 2600°C there is more C than Si in the gas phase (Figure 17.1a). In the Si-SiC-gas equilibria below 1620°C also C dominates upon Si in the gas phase (Figure 17.1b). Any corrosion of SiC by dry H_2 below 1620°C therefore is expected to remove free carbon or to generate free Si.

There are *T–X* regions, which have SiC as the only condensed phase present. In this two phase field SiC + gas a bulk composition has to be defined besides temperature and pressure. An example is shown in Figure 17.1c for the bulk composition SiC:H_2 = 1:1. At temperatures ≥1620°C C- and Si-

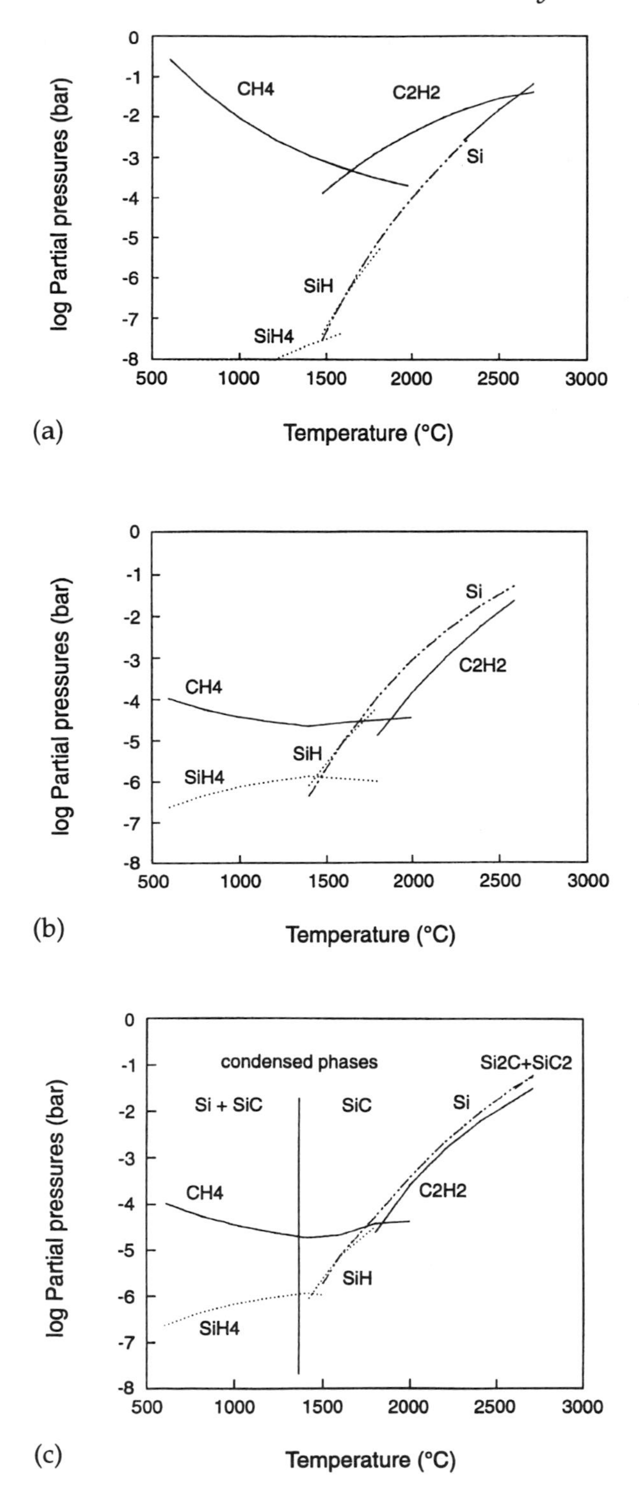

Fig. 17.1 Maximum partial pressure surface (MPPS) in the Si-C-H system for (a) SiC-C gas,(b) Si-SiC gas equilibria,(c) for a bulk composition SiC:H_2=1:1

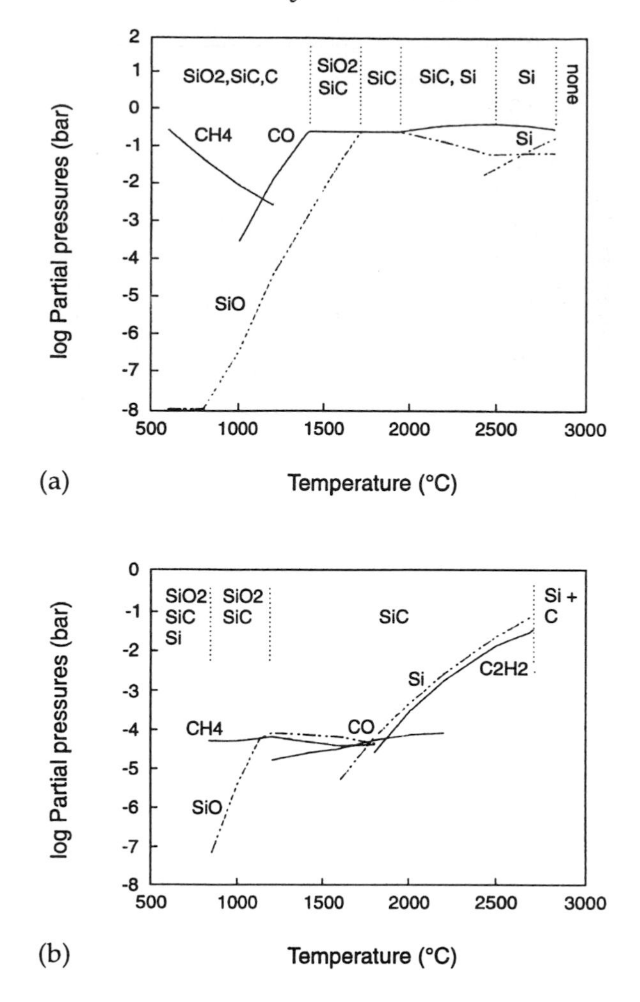

Fig. 17.2 Maximum partial pressure surface (MPPS) in the Si-C-O-H system for bulk compositions: (a) SiC:H_2O=1:1, (b) SiC:H_2:H_2O=1:1:10^{-4}. The stable condensed phases are indicated on top.

bearing species have to balance each other and a corrosion here leads to direct removal of SiC.

Si–C–O–H system

In the problem of steam or 'wet' hydrogen corrosion one of the modificationsof SiO_2 (e.g. tridymite, cristobalite or an ionic melt) may appear as additional condensed phase. The possible equilibria of interest, which are invariant under isothermal isobaric conditions are SiC–SiO_2–C gas and SiC–SiO_2–Si gas.

The invariant equilibria involving SiO_2 cannot be used for the simulation of a corrosion with the models outlined above because they imply the formation of a SiO_2-layer, which may be protective ('passivation').

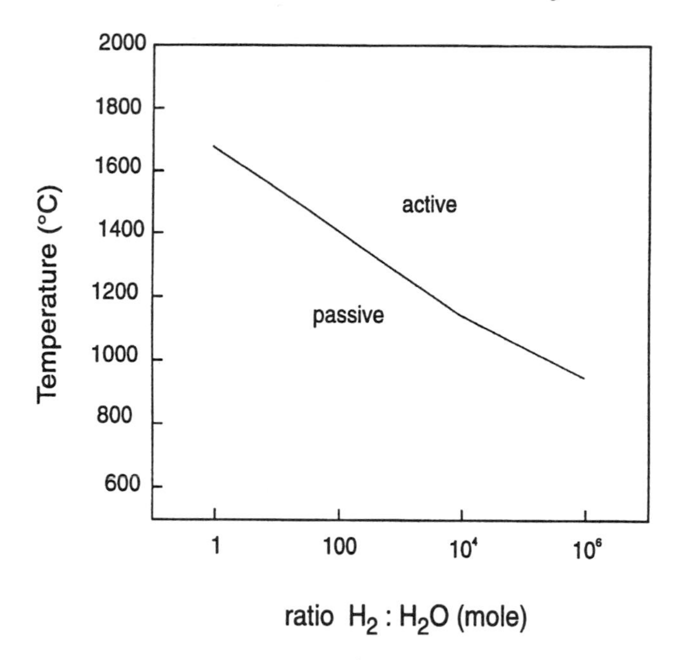

Fig. 17.3 Theoretical boundary between active and passive oxidation of pure SiC defined by the instability of SiO_2 in the phase assembled.

Outside of the four-phase equilibria we have to specify the bulk compositions to analyse the situation thermodynamically. Because all commercial SiCs have bulk compositions close to pure SiC we have calculated MPPS for pure SiC with H_2:H_2O ratios ranging from pure steam to 1ppm H_2O in H_2. Examples of this type of calculation are shown in Figure 17.2.

From Figure 17.2 it is obvious that a temperature exists, above which SiO_2 becomes unstable. This temperature may be regarded as the theoretical active/passive oxidation boundary. In three-phase equilibria in the Si–C–O–H system the transition temperature is dependent on composition. For pure SiC this transition temperature is plotted as a function of the initial H_2:H_2O ratio of the corroding gas in Figure 17.3.

Discussion

To evaluate the corrosion of SiC by hydrogen we have computed a model case, where the gas flow has a value of 1 m^3 min^{-1} through a tube of SiC with 200 mm inner diameter and 320 mm length, corresponding to an exposed surface of app. 2000 cm^2 and a gas velocity of ~ 2 km h^{-1}.

Figure 17.4 shows the predicted losses from equation (17.7) for SiC in Ar, pure H_2 and various H_2–H_2O mixtures. SiC in Ar would suffer losses in the order of mm per annum only at temperatures above 1700°C, pure Si in H_2 already at about 1500°C.

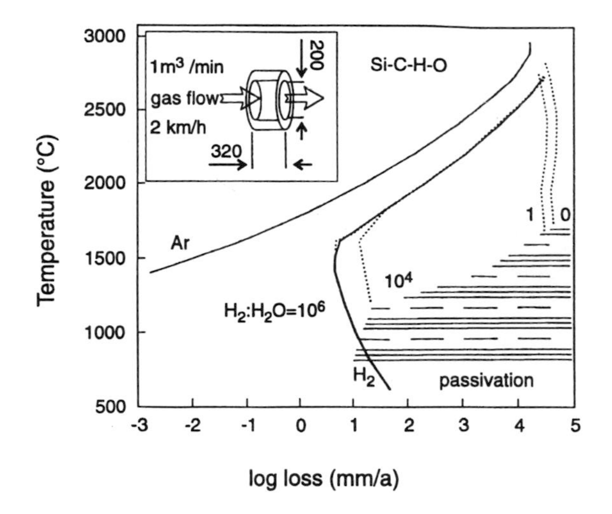

Fig. 17.4 Predicted losses of pure SiC in different environments for the model situation described by the inset drawing.

The peculiar shape of the loss-temperature curve for SiC in H_2 is a direct reflection of the appropriate MPPS (Figure 17.1), which is dominated by CH_4 up to temperatures of 1500-1600°C. Methane has a low enthalpy of formation and is thus less stable at higher temperatures. Therefore the partial pressures of CH_4 over C and SiC and the predicted losses decrease with temperature, as long as CH_4 is the dominating and hence rate-controlling gas species.

Predicted losses within the model of moderate gas flow in H_2–H_2O environments (Figure 17.4) indicate that above the passive/active transition the presence of H_2O in more than 1ppm levels increases the corrosion rate strongly.

Figure 17.5 shows the calculated influence of gas speed within this model on the corrosion rates: increasing the velocity from 2 km h^{-1} to sonic speed means a reduction in tolerable temperatures by several hundred K. The calculation of the 'worst case' gives a safe temperature limit nearly 1000°C lower.

Figure 17.2 illustrates that already quite small water additions to a system may cause the formation of a SiO_2 layer with protective potential. Furthermore it is known that SiC reacts already at room temperature with oxygen/water yielding a thin SiO_2 layer. Heating of SiC in an atmosphere containing any amount of oxygen will thus almost inevitably lead to the formation of some SiO_2. If the temperature chosen is above the active/passive transition it depends on the kinetics of the SiO_2-removal (e.g. $SiO_2 \rightarrow SiO\uparrow + {}^1/_2O_2$, $SiO_2 + H_2 \rightarrow SiO\uparrow + H_2O\uparrow$) if and when corrosion occurs. The kinetic formulations used in this paper are under the assumption of complete SiO_2-removal.

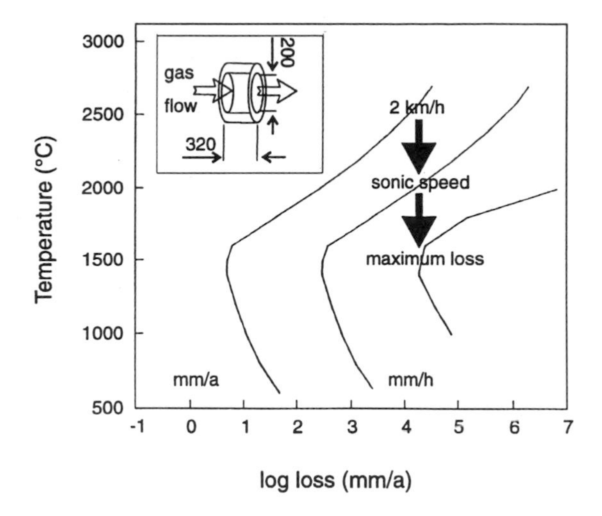

Fig. 17.5 Dependence of corrosion rate on gas velocity using the models for moderate (for 2 km/h and sonic speed) and extremely fast flow.

There are other limitations to the methods described: the kinetics of gas removal from a surface at low temperature are complex, the driving forces are strongly influenced by absorption or absorption phenomena. Hence we do not expect equation (17.7) to give good predictions at low temperatures. The transition temperatures, above which theoretical calculation and reality are in good agreement have to be determined empirically.

Hallum and Herbell [88Hal] have presented data for the behaviour of SiC under 'wet' hydrogen (H_2 with 25ppm H_2O). They reported grain boundary attack at temperatures below 1300°C followed by simultaneous SiC- and grain boundary attack to 1600°C. The investigated material was a commercial SSiC and contained randomly distributed free carbon (probably both in pockets and in grain boundaries).

From their boundary conditions (= 475 cm^3 min^{-1}, 4.86 cm^2 exposed geometrical surface) we have calculated the losses from the MPPS of the bulk composition. Hallum and Herbell [88Hal] gave estimations of partial pressures using SOLGASMIX [75Eri] type of calculations. However, they did not take into account the free carbon present, which influences the equilibrium values profoundly (compare Figures 17.1a,b, and c). We do not know the exact amount of free C in their samples and estimated it to be 0.3wt%. This crude estimate gives at random distribution of C an exposed surface of app. 0.02 cm^2.

The form equations (17.7) and (17.8) show that a higher corrosion depths X or a higher flux J is needed for a small exposed area to maintain the equilibrium pressure P_i. In the MPPSs for C-bearing SiC quite high partial pressures for CH_4 are indicated, which would effectively remove free C from pockets. However, it is known that the low temperature kinetics of CH_4

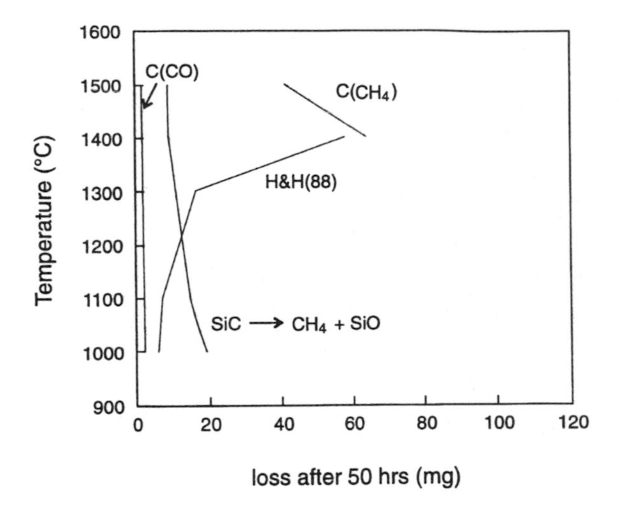

Fig. 17.6 Comparison of losses predicted from equation /17.7/ for reactions (R6), (R17) and (R21) of table 17.1 and the data of Hallum and Herbell [88Hal].

formation from C involve several steps including a transposition of CH_2 molecules, which may be slow up to temperatures of app. 1400 °C [68Gme]. The removal of C via CO is a much more effective process in the temperature range considered in the experimental study.

Because the two phases of the material are very different in their exposed surface we cannot use the bulk flux quoted by Hallum and Herbell [88Hal] for an analysis. We have rather calculated the total loss after 50 h from their data and the MPPS (Figure 17.6). The corrosion depths for C removal is large, hence a grainboundary corrosion is predicted.

We interpret their data not as an exponentional increase in loss with temperature but as a three-step feature: below 1100°C very low corrosion takes place, which is compatible with the removal of C as CO. At $T \geq 1100$ °C SiC corrosion starts via equation (R21) of table 17.1. At $T \geq 1300$°C additional CH_4 production due to increased kinetics is favoured.

This interpretation is consistent with the observed corrosion style. Thus the calculations presented in this work may be used to

- assess corrosion processes chemically (main reaction determination)
- allow extrapolation of corrosion kinetics at steady state or moderate flow conditions
- estimate save temperature limits for extreme conditions
- predict corrosion type (active/passive), predict corrosion style (grain boundary/bulk).

References

68Gme — GMELIN: *Handbuch d. Anorganischen Chemie*, 'Kohlenstoff', 8. Auflage, **B3**, Verlag Chemie, Weinheim, 1968, 795.

75Eri — G.ERIKSSON: 'Thermodynamic studies of high-temperature equilibria', *Chem. Scripta* **8**, 1975, 100-103.

82Luk — H.L.LUKAS, J.WEISS and E.T.HENIG: 'Strategies for the calculation of phase diagrams', *Calphad* **6**, 1982, 229-251.

87Ans — I.ANSARA and B.SUNDMAN: CODATA report *Computer Handling and Dissemination of Data*, 1987, 154-158.

88Hal — G.W.HALLUM and T.P.HERBELL: 'Effect of high-temperature hydrogen exposure on sintered α-SiC', *Advanced Ceramic Materials* **3**, 1988, 171-75.

18 The Carbon Potential during Heat Treatment of Steel

TORSTEN HOLM* AND JOHN ÅGREN†

Abstract

The carbon potential of a furnace atmosphere is of utmost practical importance during heat treatment of steels. The atmosphere is called active if the carbon potential differs from that of the steel and in that case there will be a transfer of carbon from the atmosphere to the steel or vice versa. This is the situation, for example, during the carburising in the case hardening process. In the other processes one rather requires an atmosphere that is inactive with respect to the steel, for example during austenitising of complex tool steels.

In order to gain a deeper understanding of the behaviour of furnace atmospheres and to make possible optimisation of heat treatment operations a cooperation was started between AGA Innovation and the Royal Institute of Technology. The project involves thermodynamic calculations of the equilibrium composition and carbon potential of furnace atmospheres used in practice and the equilibrium properties of multicomponent steels.

Introduction

The properties of all steels depend strongly on their carbon content. In metals carbon diffuses interstitially and is quite mobile even at comparatively low temperatures. The carbon exchange between furnace atmosphere and steel will thus have a large impact on the properties of the steel. For example, in case hardening one applies an active atmosphere, i.e. carbon is transferred from the atmosphere of the steel surface, resulting in a carburising. When heat treating tool steels one rather wants an inactive or inert atmosphere,i.e. there should be no transfer between atmosphere and steel. In other cases one wants a decarburising of the steel and the atmosphere must then pick up carbon from the steel surface.

The transfer of carbon between atmosphere and steel surface depends on two factors. The first one, the driving force for the transfer, is the difference in carbon activity between atmosphere and surface. The second one is the

* AGA AB Innovation, S–18181 Lidingö, Sweden
† Division of Physical Metallurgy, Royal Institute of Technology, S–10044 Stockholm, Sweden

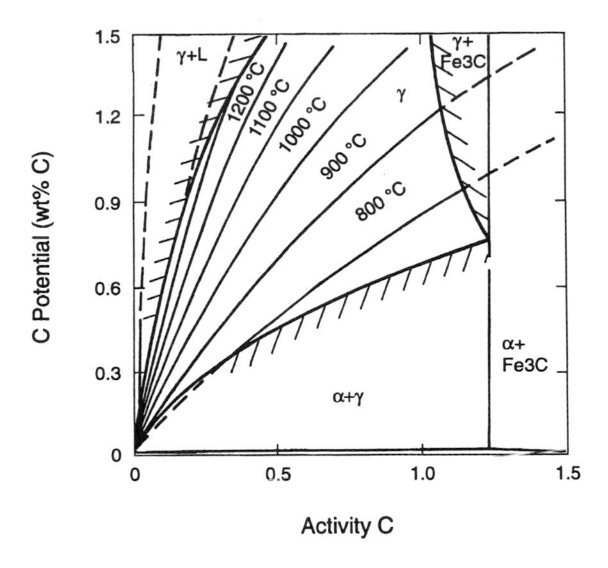

Fig. 18.1 Relation between carbon potential and carbon activity for various temperatures.

kinetics of the surface reactions. An inert atmosphere can thus be achieved in two different ways. Either one can make sure that the driving force for carbon transfer is sufficiently small or one can inhibit the surface reactions.

From the practical point of view it is thus important to be able to predict and control the carbon activity of a furnace atmosphere as well as to know the carbon activity of a particular steel.

The carbon potential

The carbon activity of an atmosphere or a steel can be predicted by equilibrium calculations. In practical heat treatment it is common to apply the so called carbon potential rather than the carbon activity. It is defined as the carbon content, expressed in weight percent, that an initially pure iron specimen would have if carbon is equilibrated between the atmosphere or the alloy under consideration.

The relation between carbon activity and carbon potential is easily calculated from the thermodynamic description of the binary Fe–C system [85Gus]. A series of such calculations have been performed on the Thermo-Calc system [85Sun] and the result is summarised in Figure 18.1. First the relation between carbon activity and carbon content of γ-iron has been calculated and plotted, one curve for each temperature. Thereafter the Fe–C phase diagram has been calculated and plotted. However, rather than using the temperature as one axis variable the carbon activity has been chosen. The carbon activity 1 corresponds to equilibrium with graphite. This freedom in choice of axis variables is build into the POLY postprocessor in Thermo-Calc.

The carbon activity in industrial furnace atmospheres

A typical furnace atmosphere for carburising consists of a mixture of H_2, CO and N_2. It has been quite common to produce this type of atmosphere by incomplete combustion of some fuel, for example propane C_3H_8, in a so called generator and then lead the products into the furnace. Such a reaction is endothermic and an atmosphere produced by this method is often called endogas. It may now be of practical interest to see how the conditions of the generator affects the constitution and more specifically the carbon potential of the resulting gas. A series of equilibrium calculations were thus performed using the Thermo-Calc program and the SGTE substance data base [87Ans]. The calculations were done as follows.

First the temperature 930°C and the pressure 1 atm(0.101325 MPa) were fixed. The overall composition given as a number of moles of the different elements were subsequently fixed. For a mixture of 1 mole of propane and x moles of air we can thus write the scheme of overall contents (assuming that the volume fraction of oxygen in air is 0.209) shown in Table 18.1.

Table 18.1 Content of elements in endogas from C_3H_8 and air.

element	content of elements in	
	1 mole C_3H_8	x mole air (0.209 O_2, N_2 balancing)
C	3	–
H	8	–
N	–	$2.x.(1 - 0.209)$
O	–	$2.x.0.209$

The above overall contents can be directly imposed as conditions in the equilibrium calculation program and with temperature and pressure fixed we will have a well posed problem with a unique solution. For example, if we consider the mixture of air and propane that yields equal amounts of C and O we will have

$$x = 3/(2.0.209) = 7.18$$

For this particular case Table 18.1 yields the following overall contents of the elements to be fixed as conditions in the Thermo-Calc program.

C	3. mole
H	8. mole
N	11.35876 mole
O	3.00124 mole

The result of such a calculation is shown in Table 18.2. The table corresponds to the output generated by the program when giving the command 'list-

Table 18.2 Equilibrium in endogas. Output from Poly-3 in Thermo-Calc.

```
Output from POLY-3, equilibrium number = 1

Conditions:
T=1203,   P=101325,   N(C)=3,   N(H)=8,   N(N)=11.35876,   N(O)=3.00124
DEGREES OF FREEDOM 0

Temperature 1203.00,   Pressure 1.013250E+05
Number of moles of components 2.53600E+01,   Mass 2.51213E-01
Total Gibbs energy -3.40940E+06,   Enthalpy 1.54050E+04,   Volume 1.24657E+00

Component  Moles          Fraction      Activity      Potential     Ref.state
C          3.0000E+00     1.4344E-01    2.2045E-01    -1.5124E+04   SER
H          8.0000E+00     3.2097E-02    6.9937E-05    -9.5702E+04   SER
N          1.1359E+01     6.3332E-01    2.1009E-06    -1.3076E+05   SER
O          3.0012E+00     1.9114E-01    7.8788E-17    -3.7089E+05   SER

GAS#1                          Status ENTERED
Number of moles 2.5360E+01,     Driving force 0.0000E+00
N 6.33322E-01   O 1.91145E-01  C 1.43436E-01   H 3.20971E-02
Constitution:
6N2        4.49718E-01   C3H6     2.97939E-12   O1       1.58724E-18
H2         3.11304E-01   C3H8     1.66545E-13   N2O1     9.94584E-19
C1O1       2.34765E-01   C3O2     1.01817E-13   C4N2     2.45047E-19
C1H4       2.00720E-03   C1N1     2.33724E-14   C1N2     8.30967E-20
H2O1       1.38672E-03   C3H6_1   2.25878E-14   C1H1     8.83721E-21
C1O2       7.57518E-04   N1O1     2.17781E-14   O2       3.57794E-21
C1H1N1     3.82008E-05   C2H1     1.23044E-14   H2O2     2.02456E-21
H3N1       2.24401E-05   C1H2     4.58264E-15   N3       2.55413E-22
C2H4       2.26349E-07   C4H4     3.27809E-15   C1       1.50360E-23
H1         1.15784E-07   C4H2     4.39020E-16   H1O2     7.02505E-24
C1H2O1     7.66874E-08   C4H6_1   3.75523E-16   H1N1O2   1.90764E-24
C2H2       7.19545E-08   C1N1α1   2.18538E-16   C3       1.33409E-24
C1H3       1.69202E-08   H1N1     1.85925E-16   N1O2     4.60848E-26
C1H1N1O1   1.25878E-08   H2N2     1.11694E-16   C2       1.01685E-26
C2H6       1.21410E-08   C2O1     8.98324E-17   H1N1O3   1.00000E-30
C4H8       1.25393E-09   C4H8_1   7.03412E-17   C5       1.00000E-30
C1H1O1     1.74495E-10   C2H4O1   2.96457E-17   C4       1.00000E-30
C1H4O1     8.15584E-11   H4N2     5.07236E-18   N1O3     1.00000E-30
H2N1       1.32726E-11   C2N1     4.56718E-18   N2O3     1.00000E-30
C2N2       5.18594E-12   H1N1O1   3.88215E-18   N2O4     1.00000E-30
H1O1       4.33422E-12   C4H10_1  3.16581E-18   O3       1.00000E-30
C3H4_1     3.94604E-12   N1       3.15136E-18
```

equilibria'. The conditions specified by the user are first listed. The overall composition corresponds to 1 mole of propane. The next set of information concerns the temperature, pressure and the elements. It should be noted that the activities listed hereare given relative and the so-called stable element reference (SER), which is the reference state used in the data bank, is used. The activity relative any other reference state is readily obtained in the program by a special command. In this case the activity of carbon relative to graphite is 1.337, i.e. there is a tendency for soot or graphite formation.

Table 18.3 shows the same type of calculation performed with 7.23 moles of air instead of 7.18. In this case the carbon activity is 0.91 and there is no tendency for formation or soot or graphite.

The above procedure can be extended in order to take into account the influence of various factors on the constitution of a furnace atmosphere. For example, if we want to study the so called synthetic atmospheres obtained from mixtures of cracked methanol CH_3OH and N_2 we can apply the same scheme as for the endogas and obtain for a mixture a cracked methanol and N_2 the overall contents shown in Table 18.4.

Table 18.3 Equilibrium in endogas. Output from Poly-3 in Thermo-Calc.

```
Output from POLY-3, equilibrium number = 1

Conditions:
T=1203,   P=101325,   N(C)=3,   N(H)=8,   N(N)=11.43186,   N(O)=3.02214
DEGREES OF FREEDOM 0

Temperature 1203.00,  Pressure 1.013250E+05
Number of moles of components 2.54600E+01,  Mass 2.52655E-01
Total Gibbs energy -3.42745E+06,  Enthalpy 1.33232E+04,  Volume 1.25209E+00

Component  Moles        Fraction     Activity     Potential    Ref.state
C          3.0000E+00   1.4262E-01   1.5007E-01   -1.8971E+04  SER
H          8.0000E+00   3.1914E-02   6.9854E-05   -9.5714E+04  SER
N          1.1438E+01   6.3409E-01   2.1036E-06   -1.3075E+05  SER
O          3.0221E+00   1.9138E-01   1.1538E-16   -3.6707E+05  SER

GAS#1                      Status ENTERED
Number of moles 2.5460E+01,   Driving force 0.0000E+00
N 6.34092E-01  O 1.91377E-01  C 1.42617E-01  H 3.19138E-02
Constitution:
N2        4.50860E-01   C3H6     9.33212E-13   N2O1     1.46017E-18
H2        3.10570E-01   C3O2     6.88763E-14   C4H10_1  6.71836E-19
C1O1      2.34030E-01   C3H8     5.20427E-14   C1N2     5.67103E-20
H2O1      2.02593E-03   N1O1     3.19325E-14   C4N2     5.27541E-20
C1H4      1.35993E-03   C1N1     1.59305E-14   O2       7.67281E-21
C1O2      1.10584E-03   C3H6_1   7.07500E-15   C1H1     6.00868E-21
C1H1N1    2.60068E-05   C2H1     5.69509E-15   H2O2     4.33137E-21
H3N1      2.23892E-05   C1H2     3.11219E-15   N3       2.56387E-22
H1        1.15647E-07   C4H4     3.42885E-16   H1O2     1.50473E-23
C2H4      1.04395E-07   C1N1O1   2.18130E-16   C1       1.02355E-23
C1H2O1    7.62668E-08   H1N1     1.85942E-16   H1N1O2   4.09125E-24
C2H2      3.32648E-08   H2N2     1.11713E-16   C3       4.20838E-25
C1H1NO1   1.25495E-08   C4H2     9.40512E-17   N1O2     9.89531E-26
C1H3      1.14774E-08   C4H6_1   8.00693E-I7   C2       4.71207E-27
C2H6      5.58637E-09   C2O1     6.09602E-17   C5       1.00000E-30
C4H8      2.66732E-10   C4H8_1   4.63703E-17   H1N1O3   1.00000E-30
C1H1O2    1.73743E-10   C2H4O1   2.00228E-17   C4       1.00000E-30
C1H4O1    8.09198E-11   H1N1O1   5.68553E-18   N1O3     1.00000E-30
H2N1      1.32581E-11   H4N2     5.06128E-18   N2O3     1.00000E-30
H1O1      6.33956E-12   N1       3.15536E-18   N2O4     1.00000E-30
C2N2      2.40925E-12   O1       2.32436E-18   On3      1.00000E-30
C3H4_1    1.23891E-12   C2N1     2.11910E-18
```

Table 18.4 Content of elements in mixtures of cracked methanol and N_2.

	content of elements in	
element	$CO + 2\,H_2$	$x\,N_2$
C	1	-
H	4	-
N	-	$2x$
O	1	-

This atmosphere will thus contain $100x/(3+x)$wt% of N_2. The conditions are given in the same way as in the previous type of calculations. The result of a series of calculations are given in Figure 18.2.

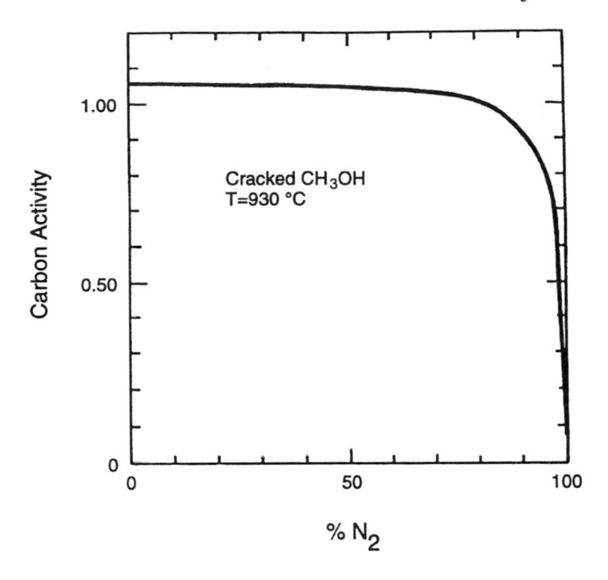

Fig. 18.2 Calculated carbon activity.

The carbon activity of multicomponent steels

As mentioned the driving force for carbon transfer between a steel and a furnace atmosphere is given by the difference in carbon activity. It is thus important to know the carbon activity of a given steel at the temperature under consideration. For a binary Fe–C steel this information can be found from Figure 18.1. For a multicomponent steel the information is conveniently accessible by means of computerised thermodynamic calculations. Let us assume for example that we want to know the carbon activity of the steel SS/AISI 2310/02 at 1000°C. The composition of this steel is shown in Table 18.5. The calculation is now performed by specifying the temperature (1273 K) and the pressure (0.101325 MPa) and the over-all alloy content in weight fraction. In addition the size of the system has to be fixed. This choice is arbitrary in the present case and we chose to consider 1 mole of atoms.

Table 18.5 Composition of steel SS/AISI 2310/02.

C	Si	Mn	Cr	Mo	V
1.5	0.30	0.45	12.00	0.80	0.90

The result of the calculation is shown in Table 18.6. As in the previous case the activities are given relative SER. The carbon activity relative graphite is 0.122. From Figure 18.1 we can see that this corresponds to the carbon potential around 0.30 wt%, i.e. a rather low carbon potential despite the high carbon content 1.5 wt% of this steel.

Table 18.6 Equilibrium in steel SS/AISI 2310/02. Output from Poly-3 in Thermo-Calc.

```
Output from POLY-3, equilibrium number = 1

Conditions:
T=1273,    P=101325,    N=1,    W(C)=1.5E-2,    W(SI)=3E-3,    W(MN)=4.5E-3,
W(CR)=1.2E-1,  W(MO)=8E-3,  W(V)=9E-3
DEGREES OF FREEDOM 0

Temperature 1273.00,  Pressure 1.013250E+05
Number of moles of components 1. 00000E+00,  Mass 5.24781E-02
Total Gibbs energy -6.52633E+04,  Enthalpy 3.48881E+04,  Volume 0.00000E+00

Component Moles        Fraction      Activity      Potential     Ref.state
VA        0.0000E+00   0.0000E+00    1.0000E+00     0.0000E+00   SER ?
C         6.5538E-02   1.5000E-02    1.8280E-02    -4.2358E+04   SER ?
CR        1.2111E-01   1.2000E-01    7.9024E-04    -7.5606E+04   SER ?
FE        7.8980E-01   8.4050E-01    2.4900E-03    -6.3458E+04   SER ?
MN        4.2985E-03   4.5000E-03    5.2795E-06    -1.2862E+05   SER ?
MO        4.3759E-03   8.0000E-03    9.1238E-05    -9.8456E+04   SER ?
SI        5.6055E-03   3.0000E-03    1.9837E-08    -1.8772E+05   SER ?
V         9.2715E-03   9.0000E-03    6.2957E-06    -1.2675E+05   SER ?

FCC_A1#1                    Status ENTERED
Number of moles 8.5332E-01,  Driving force 0.0000E+00
FE   9.12318E-01   MO   6.42898E-03   V    4.46917E-03   SI   3.38743E-03
CR   6.36060E-02   C    5.34101E-03   MN   4.44982E-03

M7C3#1                      Status  ENTERED
Number of moles 1.3803E-01,  Driving force 0.0000E+00
CR   5.86605E-01   C    8.73182E-02   MN   5.15137E-03   V    0.00000E+00
FE   2.99677E-01   MO   2.12490E-02   SI   0.00000E+00

V3C2#1                      Status ENTERED
Number of moles 8.6568E-03,  Driving force 0.0000E+00
V    8.64165E-01   SI   0.00000E+00   MN   0.00000E+00   CR   0.00000E+00
C    1.35835E-01   MO   0.00000E+00   FE   0.00000E+00
```


Summary

The Thermo-Calc program has been applied in order to investigate the behaviour of furnace atmospheres for heat treatment of steel. Particular attention is paid to the so called carbon potential. In order to control the process the carbon potential of the atmosphere must be adjusted in accordance with the carbon potential of the steel. A higher carbon potential in the atmosphere will result in carburising of the steel and a lower in decarburising.

References

85Gus   P.Gustafson: *Scand. Journ. Metall.* **14**, 1985, 259-267.

85Sun   B.Sundman, B.Jansson and J.-O.Andersson: *Calphad,* 1985, 153-190.

87Ans   I.Ansara and B.Sundman: CODATA report *Computer Handling and Dissemination of Data,* 1987, 154-158.

19 Preventing Clogging in a Continuous Casting Process

BO SUNDMAN*

Introduction

The problem presented here was caused by an attempt to modify an alloy produced by a continuous casting process. This process worked well for a stainless steel with 20 wt% Cr and the manufacturer now wanted to use the same process for a steel with 25 wt% Cr. However, he then obtained problems with clogging by solid oxide formation which prevented the flow of liquid steel.The oxide formed at the outlet was found to consist mainly of Cr_2O_3. The manufacturer thus faced an expensive and time consuming experimental scheme in order to find out how to prevent the formation of this Cr_2O_3. As an alternative route he tried to use the Thermo-Calc thermodynamic databank in order to simulate the process on the computer in order to find out a remedy. Such a simulation can usually be made in less than a day if the necessary thermodynamic data are available.

The problem is of course due to the fact that the partial pressure of oxygen, or equivalently the oxygen activity, in the liquid steel is high enough to precipitate Cr_2O_3 at the higher chromium content. The stability of Cr_2O_3 is determined by the product of the oxygen and chromium activities raised to their respective powers. Therefore the solution must be to decrease the oxygen activity in the liquid steel.

Setting up the calculation

On solving a problem of this type with a thermodynamic databank one must first try to reproduce the original process. This requires a fairly good knowledge of the conditions under which the process works. Some simplifications may be necessary if the conditions are uncertain or if there is a lack of thermodynamic data. Then one may make changes to the process by varying the conditions and in this way try to find the simplest and cheapest solution to the problem. Finally this solution should be tested by experiment before put to use. If it works one has usually saved several weeks of experimental testing of various alternative solutions. If it does not work it means one has overlooked one or more of the important factors in modelling the process.

* Division of Computational Thermodynamics, Dept. of Materials Science, Royal Institute of Technology, S–10044 Stockholm, Sweden.

The original oxygen activity was not known so a preliminary calculation was made in order to obtain an almost stable Cr_2O_3 at the original 20 wt% chromium content . The database useful for this slag+metal liquid problem is rather small so only the content of Si, Mn and Ca in the alloy and slag was taken into account together with the Cr content.
The Thermo-Calc software has more flexible ways to specify the conditions for the calculation than most other software for thermodynamic calculation. Thus one may specify the composition of an individual phase in the system as conditions and not only the overall amounts or fractions of the components. This was used to set the content of Cr, Mn and Si in the liquid metal to the values they have in the steel. The amount of CaO was set according to what had been found in the slag inclusions in the ordinary steel. In this way the oxygen activity will be fixed by the metal+slag equilibrium.

The calculation with 20 wt% Cr did not give any Cr_2O_3 which is correct according to the original alloy. The oxygen activity relative to a pure O_2 gas was $6.059.10^{-13}$ according to the calculation. In this case the activity is the same thing as the partial pressure of O_2. The alloy content of Cr was then changed to 25 wt% and a new calculation was made. The result from the second calculation showed that solid Cr_2O_3 was stable. Thus it was evident that the calculations could indeed reproduce the precipitation problems.

Solution

After satisfying oneself that the calculations are close enough to what is found in practice one may try to predict what will happen if one changes some conditions. In this case the interest was to decrease the oxygen activity which would make the Cr_2O_3 oxide less stable. All components have an effect on the oxygen activity but the question is which one would have the largest effect with the smallest change.

Thermo-Calc here offers a possibility to show, without any new calculation, the 'rate of change' or partial derivative of any state variable, for example the oxygen activity, with respect to any of the conditions that has been set for the calculation. This is done interactively by using a simple 'dot' notation as shown here:

```
command>show ac(o).w(liquid,mn)
AC(O).W(LIQUID,MN)=-8.092E-11
command>show ac(o).w(liquid,si)
AC(O).W(LIQUID,SI)=-2.543E-10
```

The symbol AC is used to denote activity for the component given within the parenthesis. The symbol W is used to denote mass fraction and this can be indexed with just a component, if one means the overall composition, or, as in this case, a phase name and a component if one means the composition

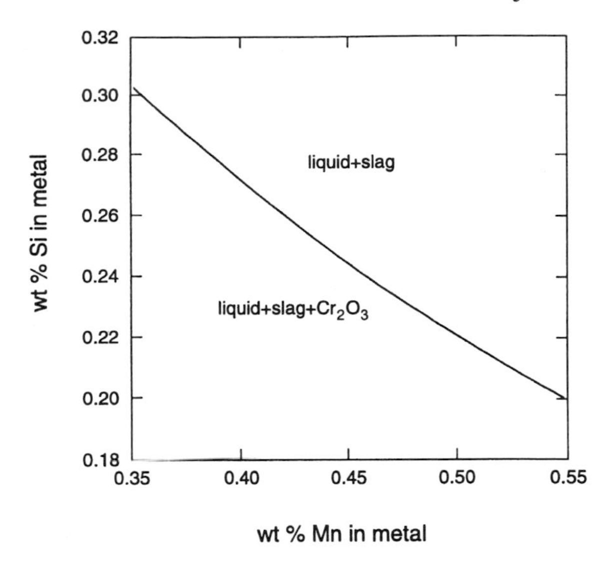

Fig. 19.1 This curve shows the solubility of Cr_2O_3<c> in liquid metal and slag for varying Si and Mn contents in the metal when the Cr-content of the liquid metal is 25 wt%.

of a phase. The negative values show that the oxygen activity will decrease by increasing the weight fraction of Mn or Si in the liquid alloy. Changing the Mn or Si content will also affect the chromium activity. This effect can be calculated in the same way

```
command>show ac(cr).w(liquid,mn)
AC(CR).W(LIQUID,MN)=5.805E-2
command>show ac(cr).w(liquid,si)
AC(CR).W(LIQUID,SI)=1.824E-1
```

This shows that an increase of the Mn or Si content will increase the chromium activity which will tend to make the Cr_2O_3 oxide more stable. However, considering that the oxygen activity is raised to 3 and the chromium activity to 2 in order to obtain the solubility product of Cr_2O_3 the overall effect will be to decrease its stability.

One may try different Mn or Si fractions manually but again Thermo-Calc offers a facility to calculate the desired answer directly. The fact one is interested in is the Mn or Si content that will make the Cr_2O_3 oxide unstable. The limiting value would be when Cr_2O_3 is just stable and one may specify this as a condition. At the same time one releases the condition on the Mn fraction and allows the program to determine this itself. The same calculation can then be repeated with the Si fraction set free.

In the present case the formation of Cr_2O_3 could be prevented by increasing the alloy content of Mn from 0.4 to 0.55 wt% or of Si from 0.2 to 0.28 wt%.

The necessary change of the Si content is smaller which is in agreement with the derivatives listed above. One may even calculate the curve giving the solubility of Cr_2O_3 for varying Mn and Si content in the liquid Metal. This is shown in Figure 19.1.

Final remarks

Many processes in steel-making are well established and if there is a problem a qualified engineer can often easily determine the cause and find a remedy. However, skilled personal is scarce and costly and it is advantageous if some of the experience that it takes years to gather from practical work can be stored into a computerised databank. Such databanks can be operated on a routine basis if it is properly adopted. The current interest in new materials for which there is little or no previous experience and the demand for less pollution and the raising cost for energy also make it necessary to develop unorthodox methods for solving manufacturing problems.

Today thermodynamic databanks like Thermo-Calc can handle most types of problems and generate almost any kind of diagrams for systems where consistent thermodynamic data are available. However, the amount of carefully assessed and consistent thermodynamic data for solution phases is very small and a collective and financially strong effort in this field is necessary.

20 Evaluation of EMF from a Potential Phase Diagram for a Quaternary System

MATS HILLERT*

Introduction

When studying the EMF of a certain electrolytic cell at 1 bar and 1000 K, one wanted to check the results by comparing with the information given in the form of potential phase diagrams. The cell had one electrode which was a mixture of MnS, MnO, Cu_2S and Cu and the other was a mixture of MnO, Mn_3O_4, Cu_2S and Cu. The electrolyte was solid zirconia stabilised with calcia.

The available potential phase diagrams are presented in Figures 20.1 and 20.2.

Theory

By assuming that there is no solubility of Cu in the Mn phases and no solubility of Mn in the Cu phases, one can construct a potential phase diagram for the quaternary Cu–Mn–O–S system by plotting all the lines in the same diagram. For a quaternary system the phase diagram at constant T and P should actually have three dimensions and the diagram now obtained is thus a projection, obtained by projecting in the μ_{Cu} direction (or the μ_{Mn} direction depending upon what potential is chosen as the dependent one). As an example, the line representing MnS+MnO in Figure 20.2 is now the projection of a planar surface, parallel with the μ_{Cu} axis and it terminates at the three-phase line MnS+MnO+Cu_2O where μ_{Cu} is so high that Cu_2S forms. All two-phase fields are two-dimensional but all the old ones only appear as one-dimensional projections, as the line representing MnS+MnO. All the new ones appear two-dimensionally and they are marked in the new diagram, Figure 20.3. All the lines are of the same kind but in order to show clearly what lines came from the Cu–O–S diagram, they are given as dashed lines.

The point of intersection between the lines representing equilibrium between MnS and MnO in Figure 20.2 and between Cu_2S and Cu in Figure 20.1 now represents a four-phase point with the four three-phase lines radiating in different directions. The one-phase fields for Cu_2S and for Cu are situated above those for MnS and MnO if the μ_{Cu} axis is plotted upwards. The points representing the two electrodes are marked with 1 and 2, respectively.

* Division of Physical Metallurgy, Royal Institute of Technology, S–10044 Stockholm, Sweden

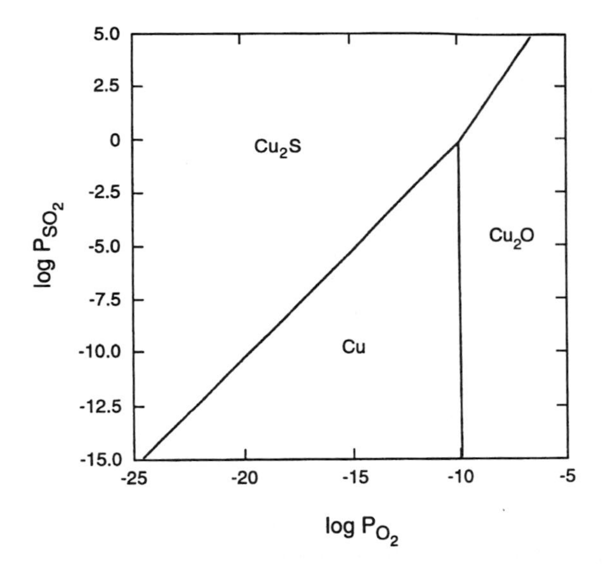

Fig. 20.1 Potential phase diagram for the Cu–O–S system at 1000 K.

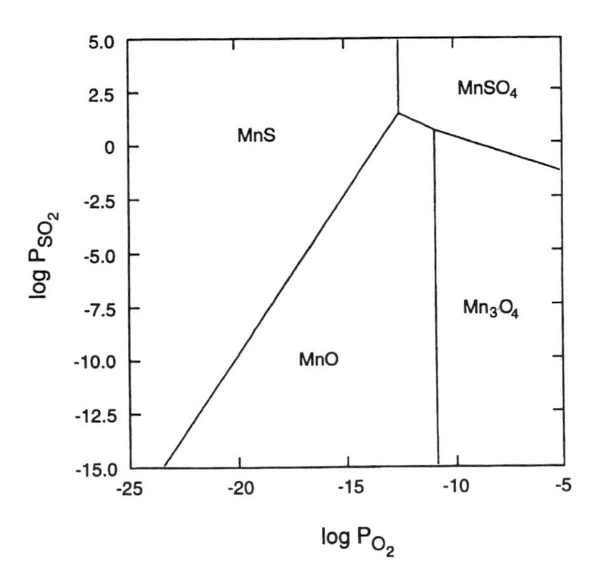

Fig. 20.2 Potential phase diagram for the Mn–O–S system at 1000 K.

The electrical current can pass through the electrolyte mainly by the diffusion of O^{-2} ions. The EMF will thus be an expression of the difference in oxygen potential between the two electrodes and it can be estimated from the difference in $RT\ln P_{O2}$ for the two points in the new diagram representing the electrodes. Because oxygen ions are divalent we obtain the EMF from $E{\cdot}2F = \Delta\mu_O$ where F is Faraday's constant (96486 coulomb/mole).

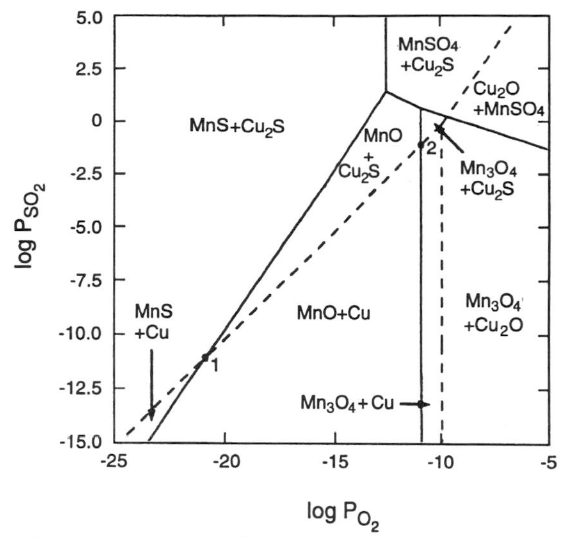

Fig. 20.3 Projection of the potential phase diagram for the Cu–Mn–O–S system at 1000 K.

Result

From the two points we read $\Delta\mu_O = 0.5\Delta\mu_{O2} = 0.5RT\ln10.(-10.9 + 20.8) = 11.4RT$ and obtain $E = 0.49$V.

The diagrams presented here were calculated from the SGTE substance database using the Thermo-Calc data bank.

Part III: Towards Process Simulation

Introduction

So far all discussions have been related to static equilibrium situations. However, all processes are dynamic in nature. One can claim that for a description of processes, thermodynamics can be applied, and that thermodynamics is a combination of thermostatics (here equilibrium thermochemistry) and kinetics. These two fields of thermodynamics are at present not in a comparable state. As was outlined in part I, thermochemistry can be used in a rigorous way. The basic theory has been transformed into computer programs, model pictures on the basis of atoms in crystal lattices or sites in molecular structures have been developed, from which Gibbs energy equations can be derived, and for many substances the necessary data are available. On the other hand, kinetics are not yet in such an advanced state, neither on the theoretical side nor regarding the data for particular substances (phases). When it comes to combining thermostatics and kinetics for simulation of processes two different approaches are presently in use, both of which are outlined in the following two chapters. Both work with the concept of local equilibrium. The major difference lies in the treatment of the flow of matter (and enthalpy).

The first method discussed treats the flow of matter and enthalpy of a macroscopic process (e.g. a shaft furnace) in a macroscopic way. No explicit kinetic data, such as diffusion coefficients or reaction rates are used. A small number of equilibrium cells is employed to describe the whole aggregate. Kinetic inhibitions can be introduced by empirical coefficients, which control the flow between the equilibrium cells. A detailed case in which the above method has been employed is given in chapter 23.

The second method treats a microscopic process (e.g. a phase transformation in an alloy) in a microscopic way. Explicit data for the diffusion coefficients of all elementary components of the alloy in all phases possible are employed in a rigorous set of equations which describe the flow of matter under the driving force of the difference in local chemical potential. A detailed case study in which the diffusion processes of a multicomponent multiphase system are treated is given in chapter 24.

21 Steady-State Calculations for Dynamic Processes

Klaus Hack

Many industrial processes are performed at high temperatures using non-isothermal furnaces into which gaseous and condensed material as well as energy are supplied at different levels, e.g. blast furnaces, reverberatory furnaces or electrothermal furnaces. Time consuming trial-and-error experiments are most often utilised for optimising such a process with regard to product yield and energy consumption. These could be made in a more systematic manner, if the temperature and composition profiles of a process could be predicted by using chemical equilibrium computations. A theoretical calculation is advantageous, because it is frequently difficult, if not impossible, to measure these profiles experimentally.

The staged reactor model approach, described below, can give a complete characterisation of a technical process. This has been demonstrated in the modelling of several alternative carbothermic silica reduction processes [78Eri]. In this approach, variations in temperature and composition inside the reactor model were predicted for various values of charge composition and enthalpy input. Conditions were then found which optimise the process investigated so as to obtain a maximum product yield at a minimum energy consumption.

General description of the reactor model

The system to be modelled is a continuously working, vertical (or horizontal) reactor into which raw material and energy can be supplied at any level. Various chemical reactions occur at different volume segments of the reactor at rates depending on the temperature, and phases formed at one level flow for further reaction at other levels, or form part of the products leaving the reactor.

To simulate such a process, the model reactor is conceptually divided into a number of sequential stages, each considered as an equilibrium cell. Inside the reactor, gaseous and condensed phases flow in opposite or parallel directions. According to these flow directions, gaseous and condensed products leave a given stage to react in neighbouring stages or leave the reactor either at opposite ends (counter-current flow) or at the same end (co-current flow).

Figures 21.1 and 21.2 show schematically the sequence of equilibrium stages for a simple 'chained' reactor for the simulation of a gas flow apparatus and for a more sophisticated counter-current aggregate, with internal gas productionrespectively.

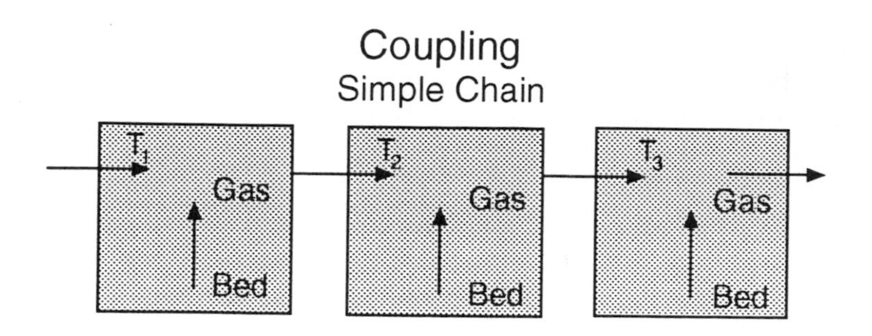

Fig. 21.1 Sequence of equilibrium stages for a simple chained reactor of a gas flow apparatus.

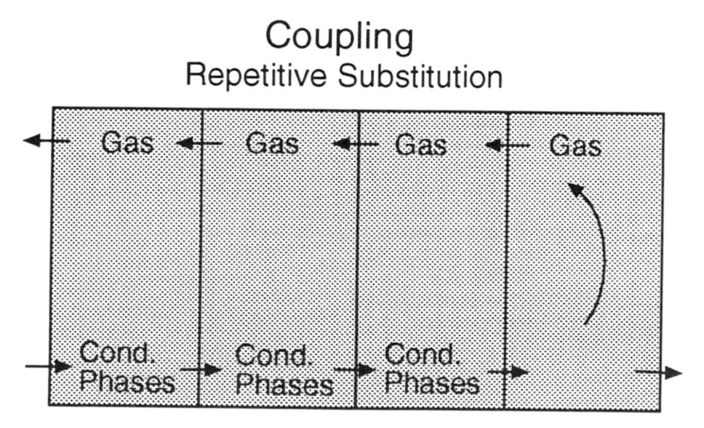

Fig. 21.2 Sequence of equilibrium stages for a more sophisticated counter-current aggregate with internal gas formation.

In order to calculate the amounts, compositions and temperatures of the flowing phases, it is assumed that the chemical reactions occurring proceed to completion. Two different types of stages are distinguished. An enthalpy-regulated stage is one for which the constant reaction temperature is determined by the chemical and thermal equilibria established. At the reaction temperature, the total enthalpy change within this stage, i.e., the sum of energies absorbed or evolved when the gaseous and condensed reactant phases are heated or cooled and the energies absorbed or evolved after the reaction is completed, counterbalances the enthalpy supplied to the stage from the outside (e.g. by electrical heating) or evolved from the stage (e.g. by heat losses). For a temperature-regulated stage the reaction temperature is fixed at the outset, and the total enthalpy of reaction can be calculated when the chemical equilibrium is known.

The solution technique employed for a counter-current reactor is one of repetitive substitution, Figure 21.2, and the iteration process may proceed in either direction throughout the reactor, from top to bottom or from bottom to top. Flows opposite to the iteration direction, calculated from a previous iteration (or initial estimate), are used as input to the next iteration, whereas flows in the other direction are calculated as the iteration proceeds. Eventually a steady-state condition will be reached, i.e., a condition characterised by a total balance in material and enthalpy input and output.

Conclusions

Based on exact thermochemical calculations and the possibility to treat kinetic constraints by a number of empirical parameters, a method is devised to model co- and counter-current flow reactors.

In addition to the set of thermochemical data (Gibbs energies) of all the phases possible in the process, the model parameters for which values must be assigned prior to the calculations are:

(1) the total number of reactor stages,
(2) amounts, compositions and temperatures of the feed phases,
(3) the stages into which the input is fed,
(4) the total enthalpy supplied to or evolved from an enthalpy-regulated stage or the reaction temperature in a temperature-regulated stage,
(5) the hydrostatic pressure of each stage,
(6) the distribution coefficients of the phases which simulate kinetic inhibitions.

Furthermore, to be able to initiate the iteration process with values of the intermediary reaction products, the reaction temperatures for the enthalpy-regulated stages and the total amounts of the elementary system components in each stage have to be estimated. For this purpose, it is useful to carry out complex equilibrium calculations for each of the reactor stages separately.

Reference

78Eri G. Eriksson and T. Johansson: *Scand. J. Metallurgy* **7**, 1978, 264–270.

22 Diffusion in Multicomponent Phases

JOHN ÅGREN*

On the microscopic scale the physical models and their mathematical counterparts available today are much more detailed and elaborate than the relatively coarse approach discussed in the previous section for macroscopic processes. Phase transformation between condensed phases for example can be treated by a method which combines an explicit description of diffusion processes with thermochemical calculations assuming local equilibrium. A short description of this method is discussed below.

Phenomenological Treatment

One dimensional diffusion along for example the z-axis in a binary system *A–B* usually obeys the well known Fick's law,

$$J_B = D_B \frac{\partial c_B}{\partial z} \tag{22.1}$$

where J_B is the diffusive flux (mol m^{-2} s^{-1}), D_B the diffusivity of B and $\partial C_B/\partial z$ the concentration gradient. If the thermodynamic behaviour of the system is sufficiently well known one may express the chemical potential of B, μ_B, for a given temperature as a function of the B concentration:

$$\mu_B = \mu_B(c_B) \tag{22.2}$$

One may thus express Fick's law in terms of the true driving force of diffusion, the gradient of μ_B, rather than the concentration gradient, i.e. [48Dar]

$$J_B = -D_B \frac{1}{d\mu_B / dc_B} \frac{\partial \mu_B}{\partial z} \tag{22.3}$$

In fact, arguments based on the theory of absolute reaction rates and principles of irreversible thermodynamics suggest that the latter expression for the flux would be a more fundamental formulation of an irreversible process. The reason is that the gradient $\partial \mu_B/\partial z$ is really the average force acting on the diffusing species. One should therefore write

$$J_B = -L_{BB} \frac{\partial \mu_B}{\partial z} \tag{22.4}$$

* Division of Physical Metallurgy, Royal Institute of Technology, S-10044 Stockholm, Sweden

and consider L_{BB} as a purely kinetic parameter indicating how easily the species moves under the action of a force. This relation may be compared with, for example, Ohm's law for the electric current. It is obvious that the L parameter plays exactly the same role as the electric conductivity. The picture can even be more generalised because one would expect a linear relationship between velocity and force when the motion of a body is opposed by the friction with the medium through which it is moving. One should thus consider L as a basic kinetic parameter characteristic for the case under consideration and by combining the last two equations one obtains

$$D_B = \frac{d\mu_B}{dc_B} L_{BB} \tag{22.5}$$

In a multicomponent alloy the chemical potential of a species normally depends on the content of all the different species and the expression for the flux of species i would be

$$J_i = -L_{ii}\frac{\partial \mu_i}{\partial z} = -L_{ii}\sum_j \frac{\partial \mu_i}{\partial c_j}\frac{\partial c_j}{\partial z} \tag{22.6}$$

One can now introduce the multicomponent diffusivity D_{ij} from the relation

$$J_i = -\sum D_{ij}\frac{\partial c_j}{\partial z} \tag{22.7}$$

where the matrix of the diffusivities D is related to the kinetic parameters L,

$$D_{ij} = \frac{\partial \mu_i}{\partial c_j} L_{ii} \tag{22.8}$$

In an ideal solution $\partial\mu_i/\partial c_j$ is different from zero only if i=j and the off-diagonal elements of the matrix formed by all the D_{ij} coefficients will vanish. However, in general chemical non-ideality prevails and **all** D_{ij} coefficients have finite values. Moreover, if the interactions are strong the off-diagonal elements may be of the same order of magnitude as the diagonal elements. A more extensive discussion of this subject is given by Kirkaldy and Young [87Kir].

Evidently, one would thus expect that the diffusion of a species does not only depend on its own concentration gradient, but also on the gradients of the other species. From a practical point of view this means that a very mobile species may be redistributed in a body if there are gradients of less mobile species even if the mobile species itself is homogeneously distributed at the

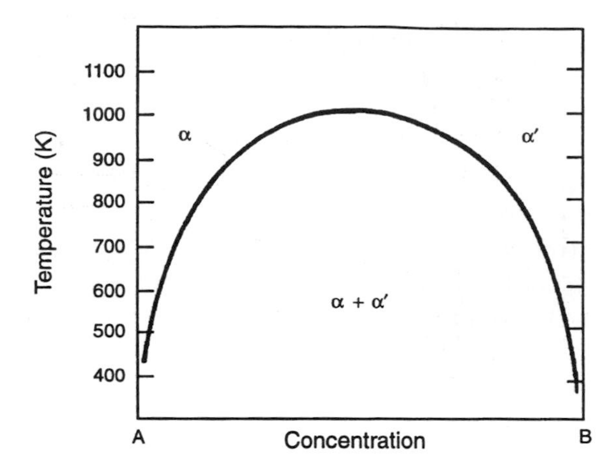

Fig. 22.1 Region of demixing.

beginning. The effect may even be so strong that an element may diffuse to regions of much higher concentration, so called up-hill diffusion. The effect was first demonstrated experimentally by Darken [49Dar]. Another evidence of the correlation of chemical potential, rather than concentration, and diffusion is given by the behaviour of a system with a miscibility gap. The region of spontaneous demixing under the spinodal curve is the region in which the chemical potential drops with increasing concentration thus driving the components into the region of higher concentration. This is demonstrated for a binary system by Figs 22.1 and 22.2.

The expression for the flux in the multicomponent case, see equation (22..6), was intended as a first approximation which excludes the possible influence of the chemical potential gradients of other species upon the flux. In general this possibility should be taken into account and one should write the flux as

$$J_i = -\sum L_{ik} \frac{\partial \mu_k}{\partial z} = -\sum L_{ik} \sum \frac{\partial \mu_k}{\partial c_j} \frac{\partial c_j}{\partial z} \tag{22.9}$$

and for the diffusivity one thus obtains

$$D_{ij} = \sum L_{ik} \frac{\partial \mu_k}{\partial c_j} \tag{22.10}$$

Analysis of experimental data – the general database

In view of the discussion above it must be concluded that in order to make the most efficient use of experimental diffusion data and to obtain reliable extrapolations one should take into account the thermochemical properties

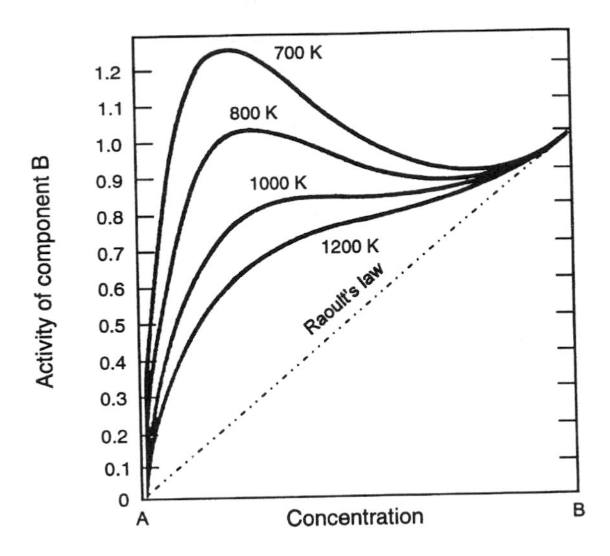

Fig. 22.2 Activities in region of demixing.

of the system when known. One should evaluate the L_{ij} parameters from diffusion experiments and store those as functions of composition and temperature in a phase-related database similar to the storage which has already be adopted for the thermochemical (Gibbs energy) data. Whenever a diffusivity is needed it can then be calculated from this coupled store of kinetic and thermochemical data. This database is in a true sense thermodynamic.

However it must be noted that detailed information on the diffusivities is still lacking in many cases and for practical calculations it may often be impossible to evaluate the off-diagonal L parameters mentioned above. It may therefore be necessary in many practical calculations to use approximations by neglecting these terms and only applying equations (22.7) and (22.8). This procedure has proven quite acceptable [88Hil]. It should also be emphasised that the simple L_{ii} parameters depend on the composition and that rather detailed experimental information is needed in order to derive a consistent set of L_{ii} parameters for a phase in a particular system.

References

48Dar L.S.Darken: *Trans. TMS-AIME* **175**, 1948, 184.

49Dar L.S.Darken: *Trans. TMS-AIME* **180**, 1949, 430–438.

87Kir J.S.Kirkaldy and D.J.Young: *Diffusion in the Condensed State*, The Institute of Metals, London, 1987.

88Hil M.Hillert and J.Ågren: *Advances in Phase Transitions*, J.D.Embury and G.R.Purdy eds,Pergamon Press, Oxford, 1988, 1–19.

23 Production of Metallurgical Grade Silicon in an Electric Arc Furnace

GUNNAR ERIKSSON* AND KLAUS HACK†

Summary

In this study, the thermochemistry of the production of metallurgical grade silicon in an electric arc furnace will be discussed in three calculational steps of increasing complexity, a stoichiometric reaction approach, a complex equilibrium calculation, and a steady state counter-current flow model.

Introduction

Metallurgical grade silicon is produced in an electric arc furnace. The process is shown schematically in Figure 23.1. Quartz sand and carbon are fed in appropriate proportions through the top and liquid silicon is extracted at the bottom. The temperature in the production zone is approximately 2200 K. This is achieved through an electric arc burning between a graphite electrode and the metal bath. Hot gases that are produced in the bottom zone of the reactor during the formation of silicon under input of energy from the electric arc. These gases flow upwards as a convective flux. On their way up, heat exchange with condensed matter falling downwards takes place. To what extent can this process be understood on the grounds of equilibrium thermodynamics?

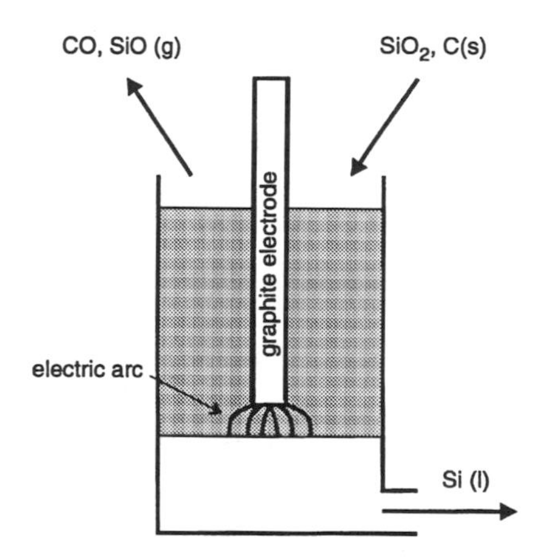

Fig. 23.1 A schematic drawing of the silicon arc furnace.

* LTH, RWTH Aachen, D–52056 Aachen, Germany
† GTT-Technologies, D–2134 Herzogenrath, Germany

The stoichiometric reaction approach

It is often claimed that the production of silicon is governed by the simple stoichiometric reaction:

$$SiO_2 + 2\,C = Si + 2\,CO\langle G\rangle \tag{23.1}$$

At equilibrium, the Gibbs energy change of the reaction must be zero. As all four phases can be considered as stoichiometric pure substances, the equilibrium constant is equal to unity. (The process takes place at atmospheric pressure, and CO is assumed to be the only gas species involved.) Thus the standard Gibbs energy change of the reaction must be zero too,

$$K = 1 \longrightarrow \Delta G^{\circ} = 0 \quad \text{where } K \text{ is the equilibrium constant} \tag{23.2}$$

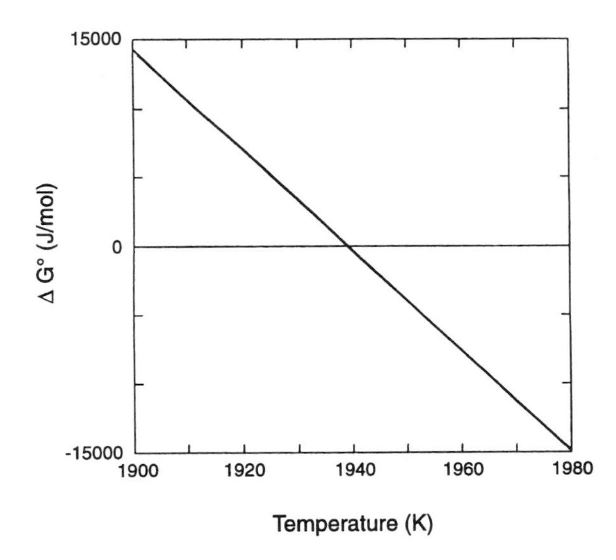

Fig. 23.2 ΔG° for the reaction $SiO_2 + 2\,C = Si + 2\,CO(g)$ as a function of T.

Figure 23.2 shows ΔG° as a function of temperature. Indeed, there is a value of T for which the curve changes sign, $T \approx 1940$ K. However, this value is far below the one known from the process. Such a difference cannot be explained by deviations from unit activities or errors in the thermodynamic data. There must be other reasons.

The complex equilibrium approach

If quartz is permitted to react freely with carbon in a system with a given total pressure and temperature, a different type of calculation must be carried out. All phases possible must be considered for set values of temperature,

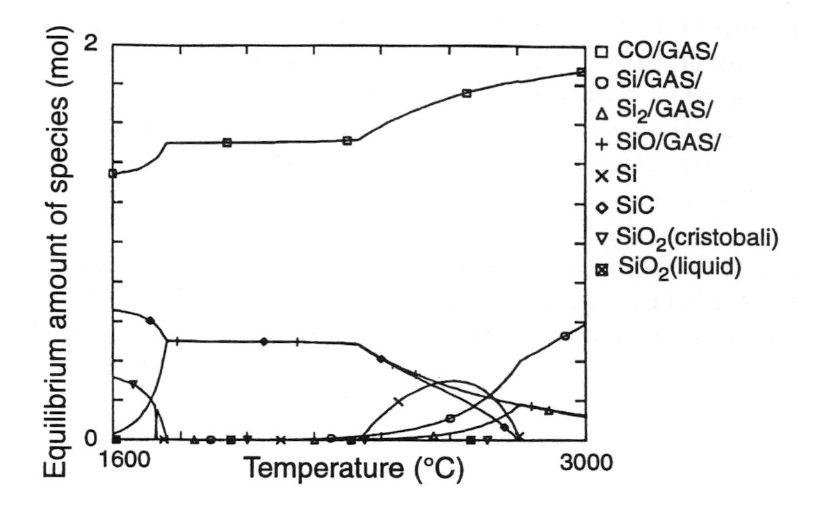

Fig. 23.3 One-dimensional phase map for 1 mol SiO_2 + 2 mol C as a function of *T*.

total pressure and system composition. Especially, all possible gas species have to be introduced into the calculation. A databank search reveals the following list of phases and phase components for the system Si–O–C :

Gas :

Si, Si_2, Si_3, SiO, SiO_2, C, O, O_2, O_3, CO, CO_2

Stoichiometric condensed:

C(graphite), SiC, SiO_2(quartz), SiO_2(tridymite), SiO_2(cristobalite), SiO_2(liquid)

Assuming stoichiometric behaviour of the reaction one mole of SiO_2 and two moles of carbon are needed as input, together with the values for the total pressure (= 1 bar) and the temperature. The complex equilibrium calculation will describe a reaction

$$SiO_2(\text{quartz}) + 2\,C \longrightarrow \eta\,\text{Gas} + \eta_{SiO2(x)}\,SiO_2(x) + \eta_{SiC}\,SiC + \eta_{Si}\,Si \quad (23.3)$$

If the temperature is varied through an interval from below the value of equilibrium for the simple stoichiometric reaction (equation (23.1)) to a value high enough to obtain liquid silicon, the resulting yield factors can be plotted as in Figure 23.3. From this figure it is obvious that the gas phase in this system contains SiO as an essential species. Thus, the simple stoichiometric reaction initially given cannot be right. On the other hand, the assumption that the process can be described as a single although complex equilibrium state is also disproved. The temperature at which silicon would be produced is near 2900 K and the yield is not more than 50 %. This is not in agreement with values known from the real process.

The counter-current reactor approach
In order to simulate the arc furnace as a whole, it is necessary to take into account the fact that the substances taking part in the process move in a temperature field while reacting. Cold condensed matter is fed through the top of the furnace, falling downwards, and hot gases flow rapidly upwards. On their way, they meet and exchange heat or even react with each other. Thus, the local mass balances must not be identical to the overall mass-balance of the process. Additionally, the temperature at different levels of the furnace is not controlled from outside, but is mainly determined by the heat exchange and the reactions taking place. Such a complex situation can only be simulated by a thermodynamic equilibrium approach if several separate zones of local equilibrium that are interconnected by materials and heat exchange are assumed.

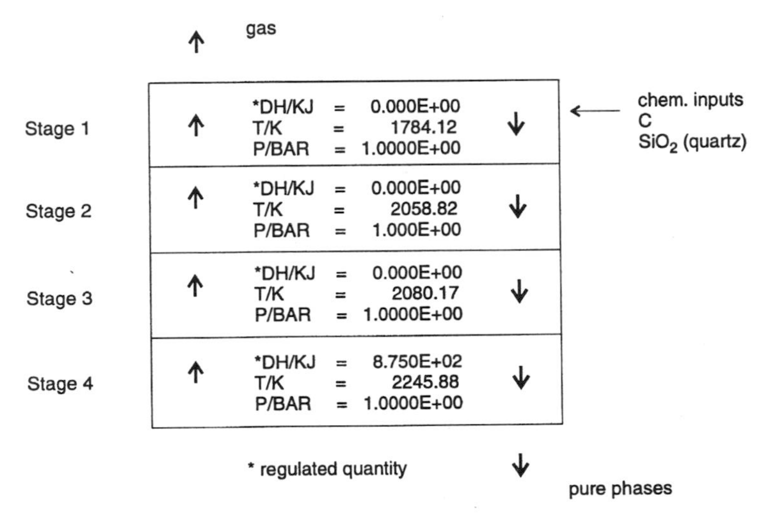

Fig. 23.4 Flow diagram for the silicon arc furnace.

For the silicon arc furnace, it was found that a reactor with four stages which are controlled by the internal heatbalance is well suited for the modelling. The flow scheme, the values for the heatbalances in each stage, the input substances, their amounts and initial temperatures as well as the distribution coefficients for the non-ideal flow between the stages are given in Figure 23.4. It should be emphasised that all values are the result of a series of parameter optimisations. This set of parameters represents best the process data obtained from the silicon arc furnace of KemaNord at Ljungaverk, Sweden.

Using the above flow data and a set of initial estimates of the elementary abundances in each stage derived from Figure 23.3, the converged solution of the reactor simulation is given by the set of Tables 23.1 to 23.4. Each of the tables represents one stage of the reactor, stage 1 being the top and stage 4

Table 23.1 Converged solution of the reactor simulation, stage 1.

STAGE 1 (ITERATION 1)

*T = 1784.12 K
P = 1.00000E+00 bar
V = 2.3002E+02 dm3

	INPUT AMOUNT mol	GAS FLOW mol	CONDENSED FLOW mol
C	2.8378E+00	1.8000E+00	1.2948E+00
O	3.8187E+00	2.0000E+00	2.2679E+00
Si	1.7810E+00	1.9988E-01	1.7733E+00

GAS	INPUT AMOUNT mol	EQUIL AMOUNT mol	MOLEFRACTION	FUGACITY bar	FLOW mol
CO	1.0377E+00	1.5428E+00	9.9494E-01	9.9494E-01	1.7999E+00
SiO	7.8081E-01	7.7238E-03	4.9810E-03	4.9810E-03	1.9985E-01
CO2	9.1223E-05	1.2212E-04	7.8751E-05	7.8751E-05	1.4525E-04
TOTAL:		1.5507E+00	1.0000E+00	1.0000E+00	

	mol	mol	ACTIVITY	mol
SiO2(cristobali)	0.0000E+00	1.1339E+00	1.0000E+00	1.1339E+00
C	1.8000E+00	6.5550E-01	1.0000E+00	6.5550E-01
SiC	0.0000E+00	6.3932E-01	1.0000E+00	6.3932E-01
SiO2(tridymite)	0.0000E+00	0.0000E+00	9.9978E-01	0.0000E+00
SiO2(liquid)	0.0000E+00	0.0000E+00	9.3374E-01	0.0000E+00
SiO2(quartz) T	1.0000E+00	0.0000E+00	9.1635E-01	0.0000E+00
Si	0.0000E+00	0.0000E+00	2.0937E-02	0.0000E+00

ENTHALPY OF REACTION = 0.0000E+00 J
ENTROPY OF REACTION = 1.1262E+02 J.K-1
THE CUTOFF LIMIT HAS BEEN SPECIFIED TO 1.000E-05
Data on 1 species identified with "T" have been extrapolated

Table 23.2 Converged solution of the reactor simulation, stage 2.

STAGE 2 (ITERATION 1)

*T = 2058.82 K
P = 1.00000E+00 bar
V = 3.0467E+02 dm3

	INPUT AMOUNT mol	GAS FLOW mol	CONDENSED FLOW mol
C	2.0703E+00	1.2948E+00	9.8154E-01
O	4.1050E+00	2.2679E+00	2.3251E+00
Si	2.8352E+00	9.7313E-01	2.1441E+00

GAS	INPUT AMOUNT mol	EQUIL AMOUNT mol	MOLEFRACTION	FUGACITY bar	FLOW mol
CO	7.7540E-01	1.0887E+00	6.1166E-01	6.1166E-01	1.2947E+00
SiO	1.0616E+00	6.9100E-01	3.8824E-01	3.8824E-01	9.7295E-01
CO2	5.7548E-05	9.9487E-05	5.5897E-05	5.5897E-05	1.1436E-04
Si	3.1171E-04	7.3605E-05	4.1355E-05	4.1355E-05	1.6745E-04
TOTAL:		1.7798E+00	1.0000E+00	1.0000E+00	

	mol	mol	ACTIVITY	mol
SiO2(liquid)	0.0000E+00	1.1626E+00	1.0000E+00	1.1626E+00
SiC	6.3932E-01	9.8154E-01	1.0000E+00	9.8154E-01
SiO2(cristobali)	1.1339E+00	0.0000E+00	9.8254E-01	0.0000E+00
SiO2(tridymite) T	0.0000E+00	0.0000E+00	9.8012E-01	0.0000E+00
SiO2(quartz) T	0.0000E+00	0.0000E+00	8.7802E-01	0.0000E+00
Si	0.0000E+00	0.0000E+00	5.0165E-01	0.0000E+00
C	6.5550E-01	0.0000E+00	1.2585E-01	0.0000E+00

ENTHALPY OF REACTION = 0.0000E+00 J
ENTROPY OF REACTION = 1.3954E+01 J.K-1
THE CUTOFF LIMIT HAS BEEN SPECIFIED TO 1.000E-05
Data on 2 species identified with "T" have been extrapolated

Table 23.3 Converged solution of the reactor simulation, stage 3.

STAGE 3 (ITERATION 1)

*T = 2080.17 K
P = 1.00000E+00 bar
V = 3.2259E+02 dm3

	INPUT AMOUNT mol	GAS FLOW mol	CONDENSED FLOW mol
C	1.7623E+00	9.8155E-01	9.7589E-01
O	4.1654E+00	2.3251E+00	2.3003E+00
Si	3.2049E+00	1.3440E+00	2.1261E+00

GAS	INPUT AMOUNT mol	EQUIL AMOUNT mol	MOLEFRACTION	FUGACITY bar	FLOW mol
SiO	1.0596E+00	1.0786E+00	5.7831E-01	5.7831E-01	1.3435E+00
CO	7.8063E-01	7.8630E-01	4.2157E-01	4.2157E-01	9.8145E-01
Si	1.0190E-03	1.5081E-04	8.0856E-05	8.0856E-05	4.0555E-04
CO2	3.0834E-05	6.4708E-05	3.4693E-05	3.4693E-05	7.2417E-05
TOTAL:		1.8652E+00	1.0000E+00	1.0000E+00	

	mol	mol	ACTIVITY	mol
SiO2(liquid)	1.1626E+00	1.1502E+00	1.0000E+00	1.1502E+00
SiC	9.8154E-01	9.7589E-01	1.0000E+00	9.7589E-01
SiO2(cristobali)	0.0000E+00	0.0000E+00	9.7691E-01	0.0000E+00
SiO2(tridymite) T	0.0000E+00	0.0000E+00	9.7388E-01	0.0000E+00
SiO2(quartz) T	0.0000E+00	0.0000E+00	8.7109E-01	0.0000E+00
Si	0.0000E+00	0.0000E+00	7.7554E-01	0.0000E+00
C	0.0000E+00	0.0000E+00	8.7603E-02	0.0000E+00

ENTHALPY OF REACTION = 0.0000E+00 J
ENTROPY OF REACTION = 2.2424E-01 J.K-1
THE CUTOFF LIMIT HAS BEEN SPECIFIED TO 1.000E-05
Data on 2 species identified with "T" have been extrapolated

Table 23.4 Converged solution of the reactor simulation, stage 4.

STAGE 4 (ITERATION 1)

*T = 2245.88 K
P = 1.00000E+00 bar
V = 4.2980E+02 dm3

	INPUT AMOUNT mol	GAS FLOW mol	CONDENSED FLOW mol
C	9.7589E-01	9.7589E-01	0.0000E+00
O	2.3003E+00	2.3003E+00	0.0000E+00
Si	2.1261E+00	1.3259E+00	8.0012E-01

GAS	INPUT AMOUNT mol	EQUIL AMOUNT mol	MOLEFRACTION	FUGACITY bar	FLOW mol
SiO	0.0000E+00	1.3245E+00	5.7544E-01	5.7544E-01	1.3245E+00
CO	0.0000E+00	9.7578E-01	4.2395E-01	4.2395E-01	9.7578E-01
Si	0.0000E+00	1.2737E-03	5.5340E-04	5.5340E-04	1.2737E-03
Si2C	0.0000E+00	6.9523E-05	3.0206E-05	3.0206E-05	6.9523E-05
CO2	0.0000E+00	3.8543E-05	1.6746E-05	1.6746E-05	3.8543E-05
Si2	0.0000E+00	2.4640E-05	1.0705E-05	1.0705E-05	2.4640E-05
TOTAL:		2.3017E+00	1.0000E+00	1.0000E+00	

	mol	mol	ACTIVITY	mol
Si	0.0000E+00	8.0012E-01	1.0000E+00	8.0012E-01
SiC	9.7589E-01	0.0000E+00	8.2082E-01	0.0000E+00
C	0.0000E+00	0.0000E+00	9.3929E-02	0.0000E+00
SiO2(liquid)	1.1502E+00	0.0000E+00	5.9571E-02	0.0000E+00
SiO2(cristobali)	0.0000E+00	0.0000E+00	5.5865E-02	0.0000E+00
SiO2(tridymite) T	0.0000E+00	0.0000E+00	5.5186E-02	0.0000E+00
SiO2(quartz) T	0.0000E+00	0.0000E+00	4.8820E-02	0.0000E+00

ENTHALPY OF REACTION = 8.7500E+05 J
ENTROPY OF REACTION = 4.1861E+02 J.K-1
THE CUTOFF LIMIT HAS BEEN SPECIFIED TO 1.000E-05
Data on 2 species identified with "T" have been extrapolated

****
CONVERGED SOLUTION OBTAINED AFTER 1 REACTOR ITERATION
****

the bottom stage. The tables contain information on the local equilibrium state (T, P and the phase amounts and compositions), the flow of matter between the stages (inflowing and outflowing substance amounts as well as elementary flows), and the boundary conditions (adiabatic behaviour (stage 1 to 3) or constant enthalpy input (stage 4)). The most important information for the derivation of a materials flow diagram of the reactor (Figure 23.5) is given in the column BOUNDARY FLOW. If an ordinate is chosen, which shows in arbitrary units the width and sequence of the four stages, the calculated materials flow can be marked for the upflowing gases on the upper abscissa in each segment, and for the downflowing condensed phases on the lower abscissa in each segment.

Thus, continuous curves are obtained which represent the total materials flow through the reactor relative to one mole of silica and 1.8 mol of carbon in the feed. The most important result, the silicon yield, can be read on the lowest abscissa to be 0.8 moles per mole of silica. The total elementary mass balance of the reactor (Table 23.5) is readily obtained from the values on the outermost abscissae.

As a mass balance with respect to phases on obtains:

$$SiO_2 + 1.8\,C \longrightarrow 0.8\,Si + 1.8\,CO\text{<G>} + 0.2\,SiO\text{<G>} \quad (23.4)$$

Note that this is not an isothermal reaction equation as silica and carbon enter the reactor at room temperature, silicon leaves the reactor at 2200 K and the gasphase leaves at 1874 K.

The temperature distribution that results from the energy input in the production zone (ΔH = +875 kJ/mol SiO_2) and the assumed adiabatic behaviour ($\Delta H = 0$) of the three upper zones is also given in the diagram. The highest temperature is reached in the production zone. Its value (2245 K) is higher than that for which the standard Gibbs energy change of the simple reaction in the first paragraph changes sign (1940 K); but, it is also considerably lower than the value for which the single cell complex equilibrium calculation shows a maximum of the silicon yield (appr. 2900 K).

A thermochemical understanding of the different reactions taking place at different levels within the reactor can be obtained from the materials flow between stages 3 and 4 and stages 1, 2 and 3, respectively. The reactor can obviously be split into two distinct zones that are governed by two separate processes. In the bottom zone, silicon is produced according to the mass balance equation

$$SiO_2 + SiC \longrightarrow Si + SiO\text{<G>} + CO\text{<G>} \quad (23.5)$$

In the top zone, the upflowing silicon monoxide reacts with the carbon according to the mass balance equation

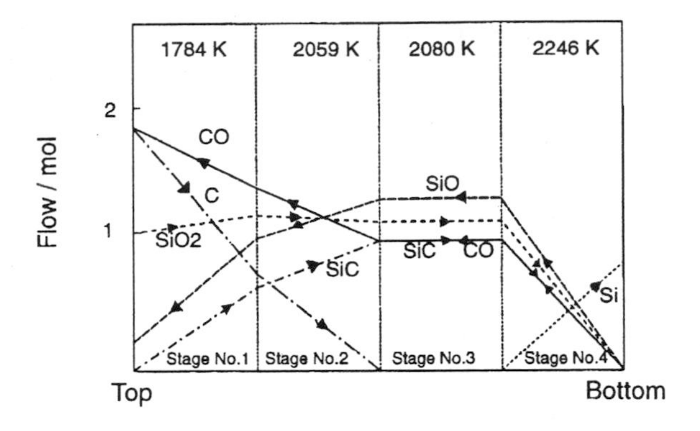

Fig. 23.5 The materials flow for the calculated steady state.

Table 23.5 Total elementary massbalance of the reactor.

Component	Input [mol]	Output [mol]
Si	1 SiO_2	0.8 Si + 0.2 SiO<G>
C	1.8 C	1.8 CO<G>
O	2 SiO_2	1.8 CO<G> + 0.2 SiO<G>

$$C + SiO\langle G\rangle \longrightarrow {}^{2}/_{3}SiO_2 + {}^{1}/_{3}SiC + {}^{2}/_{3}CO\langle G\rangle \quad (23.6)$$

This reaction is, however, usually incomplete because the amount of incoming carbon is too low to allow all SiO<G> to react. Thus, a loss of silicon (20%) cannot be avoided.

It is worth noting that SiC is both formed and consumed within the reactor and, therefore, does not occur in the total mass-balance. However it is an essential phase in the whole process!

A more detailed analysis of the process was given by Eriksson and Johansson [78Eri].They have also employed modifications of the above set of calculational parameters, e.g. to study the influence of the energy supply and the amount of carbon fed into the process. Figure 23.6 shows the silicon yield as a function of energy supply for two different amounts of carbon feed. The lower curve (1.8 mol carbon) shows that the optimum value for the silicon yield (80%) is obtained for 875 kJ per mole of silica. This value was used in the calculations discussed above. A higher energy supply will not raise the silicon yield. For a higher carbon feed (2 mol per mol of silica), the upper curve indicates a possible silicon yield of more than 95%. However, the resulting temperature level in the bottom stage (>2700 K) would make such a combination of process parameters technologically unfeasible.

The calculations for this case study have been performed using ChemSage.

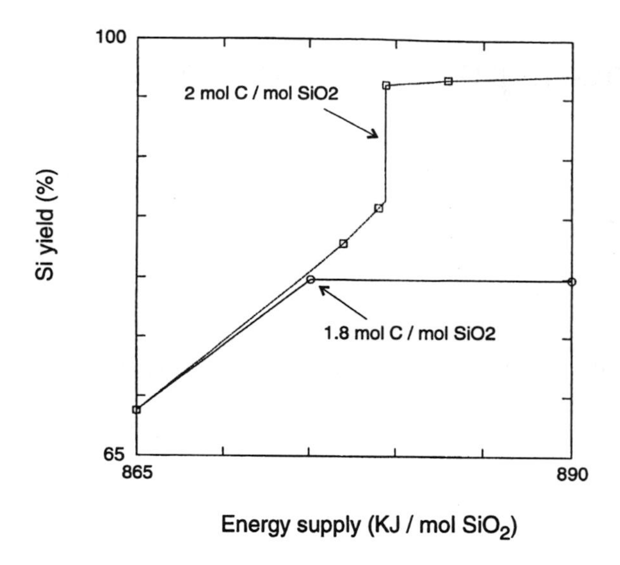

Fig. 23.6 The silicon yield for two different amounts of carbon feed.

Reference

78Eri G.ERIKSSON and T. JOHANSSON: *Scand. J. Metallurgy* 7, 1978, 264–270.

24 Multicomponent Diffusion in Compound Steel

JOHN ÅGREN*

Abstract

If the different components in a compound material are chemically incompatible the heat treatment of such a material will present some difficulties. There may be strong driving forces for the transfer of atoms by means of diffusion between the two component materials. This transfer may result in a drastic change in the properties.

Here we will present a detailed study of diffusion in a compound steel by means of numerical solution of the diffusion equations.

We will also present a more approximate calculation where the mobile carbon is equilibrated between the steels and the less mobile substitutional elements are only equilibrated within each steel but not between steels. Thus, complete equilibrium is established within each steel. In the latter calculation both steels are multicomponent and multiphase materials.

Introduction

Compound materials are increasingly being used as structural materials. For example, one can combine a corrosion resistant steel with an inexpensive low-alloy high-strength steel and thereby lower the costs. However, if the different materials are chemically incompatible the heat treatment of such a compound material will present some difficulties. There may be strong driving forces for the transfer of atoms by means of diffusion between the two component materials. This transfer may result in a drastic change in the properties close to the interface between the two materials. In some cases this may be beneficial but in most cases it is not and the amount of transfer must be minimised.

Over the last decade a general package of computer programs for multicomponent diffusional transformations [82Agr] has been developed and the purpose of the present report is to describe its application to some problems of practical interest.

Numerical calculation of diffusion between a stainless steel and a tempering steel

In this section we consider the heat treatment of a compound material which is composed of a 16Cr-10Ni austenitic stainless steel (A) and a 1Cr-

* Division of Physical Metallurgy, Royal Institute of Technology, S–10044 Stockholm, Sweden

Table 24.1 Alloy composition in weight percent.

Steel	C	Cr	Mn	Mo	Ni	Si	V	W
A	0.025	16.5	0.65	2.08	10.3	0.66	–	–
B	0.31	1.12	0.43	0.05	4.21	0.26	–	–
C	0.40	–	0.30	–	–	0.30	–	–
D	0.95	4.00	0.30	5.00	–	0.30	2.00	6.30

4Ni tempering steel (B). The material is heat treated in two steps, 2 h at 1250°C and 30 min at 900°C. The important question now is to what extent the diffusional reactions occur. Further, we want to know if the reactions can be inhibited by introducing a thin sheet of pure nickel between the steels.

The complete heat treatment cycle was now simulated under various conditions with Ågren's [82Agr] program package. The full chemical compositions given in Table 24.1 were entered as initial condition in the two parts of the compound material and the program calculated the subsequent development of the concentration profiles. The thermodynamic data compiled by Uhrenius [78Uhr] and the diffusivities compiled by Fridberg et al. [69Fri] were applied. It was assumed that both steels are one-phase austenitic during the heat treatment. The results plotted by the computer are shown in Figures 24.1 - 24.8 and will now be discussed. The stainless steel denoted A is at the left side in all diagrams.

Figure 24.1 shows the carbon-concentration profile at the end of the heat treatment. As can be seen there is a considerable exchange of carbon between the two steels. The affected zone ranges over several mm and there is a very drastic change close to the interface. The profiles of all the components except Ni are shown in Figure 24.2 with a finer scale on the x axis. The profiles of Cr, Mo, Mn and Ni (not seen in the diagram) have more or less the step behaviour whereas the Si profile shows a complex variation similar to that of carbon. The behaviour is caused by the fact that C diffuses several orders of magnitude faster than the other elements. The Si profile has been plotted separately in Figure 24.3.

The effect of introducing a 100 μm layer of pure Ni between the two steels was now investigated by the same procedure. The resulting carbon concentration profile is shown in Figure 24.4, solid line. As a comparison the profile without Ni layer, i.e. Figure 24.1, has been superimposed, dashed line. As can be seen the carbon redistribution is inhibited to some degree by the Ni layer.

Calculation of partial equilibrium between a carbon steel and an alloy steel

In this section we will consider the partial equilibrium with respect to carbon between a carbon steel and an alloy tool steel. The full compositions are

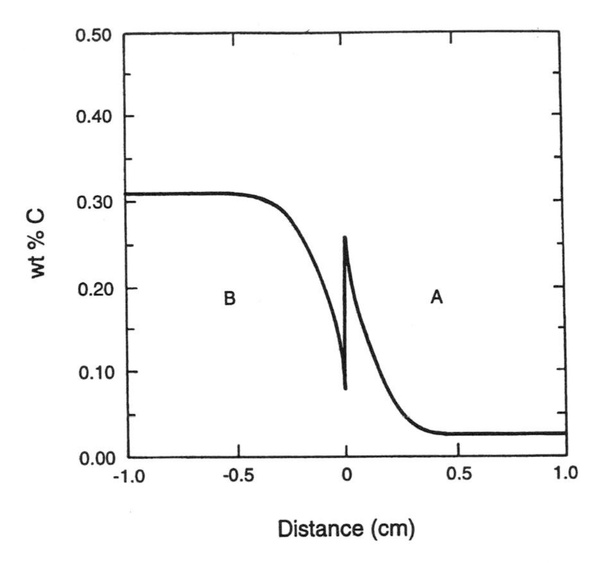

Fig. 24.1 Calculated carbon concentration profile in compound steel after 2h at 1250°C and 30 min at 900°C.

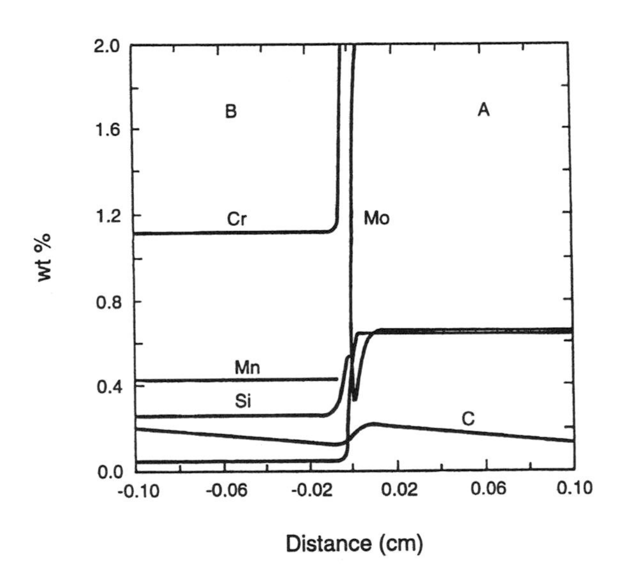

Fig. 24.2 Calculated concentration profiles of Cr, Mn, Ni, Si and C in compound steel after 2h at 1250°C and 30 min at 900°C.

given in Table 24.1, steel C and D. The calculation should be a reasonable approximation if the temperature is low enough to neglect the redistribution of substitutional elements between the steels but high enough to allow a complete equilibration of carbon. The calculation is performed by

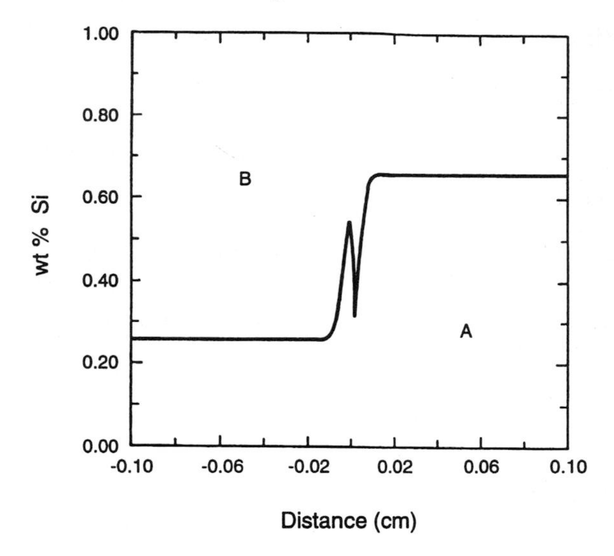

Fig. 24.3 Calculated Si concentration profile in compound steel after 2h at 1250°C and 30 min at 900°C.

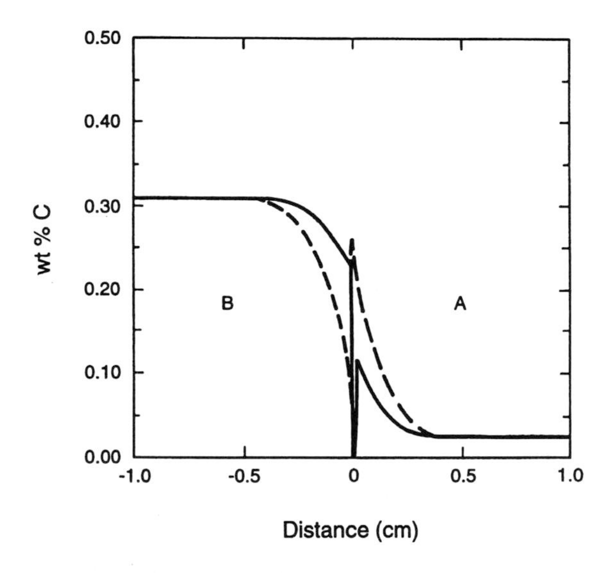

Fig. 24.4 Calculated carbon concentration profile in compound steel with 100μm Ni layer after 2h at 1250°C and 30 min at 900°C, solid line. Corresponding carbon concentration profile without Ni layer, see Figure 24.1, is superimposed, dashed line.

defining two separate equilibria having the same carbon activity. The two equilibria are connected by the condition that the total number of carbon atoms must be equal to the sum of the initial carbon contents. The calculation is performed by means of the PARROT program developed by Jansson

Table 24.2 Listing from the Parrot program for equilibrium in the tool steel at 600 °C (873K).

```
Output from POLY-3, equilibrium number = 2

Conditions:
P=P0,   T=T0,   AC(C)=AC1,   B(SI)=3E-1,   B(MN)=3E-1,   B(CR)=4,   B(MO)=5,
B(V)=2,  B(W)=6.3,  B(FE)=81.25
DEGREES OF FREEDOM 0

Temperature 873.00,  Pressure 1.013250E+05
Number of moles of components 1.78593E+00,  Mass 1.00499E-01
Total  Gibbs  energy  -1.32465E+04,    Enthalpy  -1.70632E+04,      Volume
0.00000E+00

Component Moles        Fraction     Activity     Potential    Ref.state
VA        0.0000E+00   0.0000E+00   1.0000E+00   0.0000E+00   SER
C         1.1235E-01   1.3427E-02   2.8640E-01  -9.0759E+03   SER
CR        7.6929E-02   3.9801E-02   5.7437E-02  -2.0738E+04   SER
CU        0.0000E+00   0.0000E+00   1.0000E+00   0.0000E+00   SER
FE        1.4549E+00   8.0846E-01   8.6128E-01  -1.0840E+03   SER
MN        5.4607E-03   2.9851E-03   2.2078E-03  -4.4392E+04   SER
MO        5.2116E-02   4.9752E-02   8.3779E-03  -3.4712E+04   SER
NI        0.0000E+00   0.0000E+00   1.0000E+00   0.0000E+00   SER
SI        1.0682E-02   2.9851E-03   1.4399E-09  -1.4778E+05   SER
V         3.9261E-02   1.9901E-02   5.3354E-07  -1.0484E+05   SER
W         3.4267E-02   6.2687E-02   5.1939E-03  -3.8182E+04   SER

FERRITE#1                      Status ENTERED
Number of moles 1.3579E+00,         Driving force 0.00000E+00
FE   9.87117E-01   MN   1.07526E-03   V    1.07362E-05   NI    0.00000E+00
CR   6.36745E-03   MO   8.90227E-04   C    6.24585E-06
SI   3.97066E-03   W    5.62529E-04   CU   0.00000E+00

M23C6#1                        Status ENTERED
Number of moles 1.2839E-01,         Driving force 0.0000E+00
FE   4.38725E-01   C    5.06997E-02   NI   0.00000E+00   SI    0.00000E+00
CR   3.19441E-01   MN   3.39701E-02   V    0.00000E+00
MO   1.42116E-01   W    1.50474E-02   CU   0.00000E+00

M6C#1                          Status ENTERED
Number of moles 1.5889E-01,         Driving force 0.00000E+00
W    4.24452E-01   C    2.05929E-02   MN   0.00000E+00   SI    0.00000E+00
FE   2.72052E-01   CR   1.79603E-02   V    0.00000E+00
MO   2.64943E-01   NI   0.00000E+00   CU   0.00000E+00

M7C3#1                         Status ENTERED
Number of moles 3.6774E-02,         Driving force 0.00000E+00
CR   7.02834E-01   MN   3.11339E-03   NI   0.00000E+00   W     0.00000E+00
FE   2.05261F-01   SI   0.00000E+00   V    0.00000E+00
C    8.87913E-02   MO   0.00000E+00   CU   0.00000E+00

MC_FCC_CARBIDE#1               Status ENTERED
Number of moles 1.0402E-01,         Driving force 0.00000E+00
V    5.09932E-01   MO   1.35373E-01   FE   4.36746E-05   NI    0.00000E+00
C    1.59345E-01   CR   5.66387E-02   SI   0.00000E+00
W    1.38580E-01   MN   8.78542E-05   CU   0.00000E+00
```

[84Jan]. Two equilibria are defined. Both equilibria are defined by fixing a common temperature, pressure and carbon activity. In addition the appropriate alloy contents of the individual steels are fixed. However, the carbon activity is not known but must be chosen in such a way that the sum of the carbon contents must equal a fixed value. In principle this can be done by trial and error but the Parrot program allows this condition to be added as an extra constraint and the unknown carbon activity to be obtained from an optimisation.

A calculation was now performed for a compound material consisting of equal weights of a carbon steel and a tool steel of type M2. The calculation was performed for T=600 °C. The initial carbon contents are 0.4 wt% C in the carbon steel and 0.95 wt% C in the tool steel. The corresponding carbon activities calculated from the data by Uhrenius [78Uhr] at 600°C are 2.05 and 0.02 respectively. Despite the much lower carbon content there is thus a strong tendency for carbon to diffuse from steel C to D. The final result yields the common carbon activity 0.29 and almost all the carbon redistributed to the tool steel. Table 24.2 shows the final equilibrium of steel D as listed in the Parrot program. As can be seen 4 different carbides and a ferritic matrix coexist at equilibrium.

Summary

It has been shown that diffusion calculations and modified equilibrium calculations can give valuable information for the heat treating of compound materials. As input the diffusivities of the various components and a thermodynamic description are required. A strong redistribution of carbon is predicted. In order to design a proper heat treatment of such a compound material it is absolutely necessary to take this carbon redistribution into account.

References

82Agr — See for example: *J.Phys.Chem.Solids* **43**, 1982, 385 or *Acta Metall.* **30**, 1982, 841.

78Uhr — B.UHRENIUS: in *Hardenability Concepts with Applications to Steel* (D.V.Doane and J.S.Kirkaldy eds) TMS-AIME, 1978, 28.

69Fri — JFRIDBERG, L.-E.Törndahl,L.-E. and M.Hillert: *Jernkont. Ann.* **153**, 1969, 263.

84Jan — B.JANSSON: TRITA-MAC-0234, 1984, Internal report. Division of Physical Metallurgy, Royal Inst. of Techn. S-100 44 Stockholm, Sweden.

Name Index

Adam, G. 18
Ågren, John 39, 101, 139
on carbon potential during heat treatment of steel 176-82
on diffusion in compound steel 209-14
on diffusion in multicomponent phases 196-9
Alcock, C.B. 13, 55
Andersson, J.-O. 14, 92, 98, 177
Ansara, Ibrahim 105, 106, 138, 158, 167, 178
on solidification paths for multi-component system 94-8
Arena, C. 116

Bale, C.W. 14, 57
Ball, Richard G.J.: on nuclear reactor accidents 135-50
Barbier, J.N. 111, 113
Barin, I. 14
Barry, Tom I. 14, 139
on Cu-Ni-Fe sulphide ores 151-62
on hot salt corrosion of superalloys 56-69
Bernard, Claude 53
on CVD of WSi_2 108-17
Bernstein, H. 14
Besmann, T.M. 115
Bhansali, A.S. 139
Blander, M. 40, 158
Blanquet, E. 111, 113
Bramblett, T.R. 43
Butland, A.T.D. 137

Chang 155
Chatterji, N. 103
Chevalier, P.Y. 111, 139
Cheynet, B. 14
Choudray 155
Christ, CH.L. 119
Cole, R.K. 143
Cordfunke, E.H.P. 138

Darken, L.S. 196, 198
Davies, R.H. 14, 111
Debye, P. 27
Deniel, Y. 111
Dinsdale, Alan T. 14, 53, 106
on Cu-Ni-Fe sulphide ores 151-62
on hot salt corrosion of superalloys 56-69
Doane, D.V. 210, 214
Ducarroir, M.J. 111
Duchemin, J.P. 115

Einstein, Albert 27
Ellis, M.A. 143
Eriksson, Gunnar 14, 53, 111, 122, 167, 173, 193
on silicon production in electric arc furnace 200-8
Evans, W.H. 122

Fei, Y. 103
Fernández Guillermet, Armando 101, 158
on cemented WC tools 77-84
Ford, M.J. 77
Frassek, B. 14
Fridberg, L.-E. 210
Frohberg, G.M. 40, 158

Gallagher, R. 14
Garrels, R.M. 119
Gaye, H. 40, 158
Geisler, J. 131
Gerthsen, C. 28

Gibbs, J.W. 17, 105
see also Gibbs energies *in subject index*
Gisby, J. 14, 139, 158
Grade, K. 131
Gupta, D.K. 67
Gupta, H. 43,
Gurland, J. 77
Gustafson, Per 177
on computer assisted development of steels 70-6

Hack, Klaus 53
on basic relationships 17-24
on Fe and Cu ions in aqueous H_2SO_4 118-28
on graphical representation of equilibria 43-50
on hot isostatic pressing of Al-Ni alloys 103-7
on low carbon steels 129-34
on models and data 25-42
on silicon production in electric arc furnace 200-8
on steady-state calculation 193-5
on summarising mathematical relationships 51-2
Hallstedt, B. 92, 101
Hallum, G.W. 173, 174
Harvie, C.E. 53
Henig, E.T. 139, 167
Herbell, T.P. 173, 174
Hillert, Mats 39, 105, 139, 158, 199, 210
on EMF evaluation 187-92
on prediction of loss of corrosion resistance 85-93
on prediction of quasiternary section 99-102
Hittmair, O. 18
Hodson, Susan M. 14, 111
on Cu-Ni-Fe sulphide ores 151-62
Höglund, L. 92
Holleck, H. 77
Holm, Torsten: on carbon potential during heat treatment of steel 176-82
Horrocks, P.J. 140
Humenik, M.Jr 77

Inden, G. 29

Jacquot, A. 111
Jansson, B. 14, 39, 92, 98, 101, 139, 177, 212-13
Johansson, T. 193, 207
Jonsson, Stefan: on prediction of quasiternary section 99-102

Kapoor, M.L. 40, 158
Kaufman, L. 14
Kelley, D.P. 143
Kelley, K.K. 28
Kieffer, R. 77
Kirkaldy, J.S. 197, 210, 214
Kister, A.T. 33, 40
Klein, R. 116
Kneser, H.O. 28
Konings, R.J.M. 138
Kowalski, M. 131
Korb, Jürgen: on Fe and Cu ions in aqueous H_2SO_4 118-28
Kubaschewski, O. 13, 55

Lacy, M. 14
Landau, A.I. 43, 44
Lang, C. 105, 106
Laqua, W. 67
Liang, W.W. 57
Lin, P.L. 57
Lindenberg, H.-U. 132
Lukas, Hans L. 139, 140
on high-temperature corrosion of SiC 163-75
Luthra, K.L. 67
Lyhs, W. 53

Mackey 155
Madar, Roland 53
on CVD of WSi_2 108-17
Mallik, A.K. 139
Mariaux, S. 116
Masing, G. 43, 44
Mason, Paul K.: on nuclear reactor accidents 135-50
Mastromatteo, E. 116
Meintjes, K. 53
Mignanelli, Mike A.: on nuclear reactor accidents 135-50
Million-Brodaz, J.F. 109
Moeller, N. 53
Morgan, A. 53
Morral, J.E. 43
Moskowitz, D. 77
Mucklejohn, S.A. 53
Murnaghan, F.D. 31, 103
Murray, J. 103, 105, 106

Nagamori 155

Nesor, H. 13
Neumann 155
Neuschütz, D. 131
Nickel, Klaus G.: on high-temperature corrosion of SiC 163-75
Nishida, K. 67
Nowotny, H. 43

Palatnik, L.S. 43, 44
Pawlek, F. 119
Pelton, A.D. 14, 40, 43, 57, 104, 158
Petzow, Günter: on high-temperature corrosion of SiC 163-75
Poirier, J.P. 53
Pons, M. 116
Potter, P.E. 53, 137, 138
Prakash, L. 77
Prince, A. 43, 44, 46
Prins, G. 138
Pugh, N.J. 14

Qiu, Caian: on prediction of loss of corrosion resistance 85-93

Rand, M.H. 53, 138, 140
Rapp, R.A. 67
Rauscher, K. 121
Redlich, O. 33, 40
Roberts, G.J. 137
Roche, M.F. 143
Roine, A. 14

Saxena, S.K. 53, 103
Schmalzried, H. 43, 67, 104
Schnedler, E. 14
Schubert, K.-H. 132
Schwartzkopf, P.S. 77
Shen, G. 103
Shores, D.A. 67
Smith, P.N. 137
Sommer, F. 139
Spear, K.E. 115
Spencer, P.J. 14, 55, 131
Staffansson, L.-I. 101
Stawström, C.O. 86
Sundman, Bo 14, 16, 39, 92, 101, 105, 106, 138, 139, 158, 167, 177, 178
on clogging prevention in continuous casting 183-6
on solidification paths for multi-component system 94-8

Taylor, Jeff R. 14
on Cu-Ni-Fe sulphide ores 151-62
Temkin, M. 37
Thomas, N. 113
Thompson, W.T. 14
Thümmler, F. 77
Törndahl, L.-E. 210
Touloukian, Y. 105, 106
Turnbull, A.G. 14

Uhrenius, B. 210, 214

Vahlas, Constantin: on CVD of WSi_2 108-17
Valentin, P. 131
Vay, P. 111
Voigt, J. 121

Wadsley, M.W. 14
Wagman, D.D. 122
Weare, J.H. 53
Weiss, J. 167
Welfringer, J. 40, 158
Wilhelmi, H. 53
Wilke, I and K.-TH. 121
Wisell, H. 73, 74

Young, D.J. 197

Zimmerman, B. 139
Zörcher, Z. 132

Subject Index

Notes: 1. Symbols for chemical elements and compounds are used, instead of full names. 2. Names of people are in the separate name index. In some cases where their names have been given to functions eg Gibbs energies, they are included in this subject index

Ag-In-Cd system 140
Al 146
Al-Mg-Si solidification paths 94-8
Al-Ni alloys *see under* hot isostatic pressing
and CVD process 108
and hot salt corrosion of superalloys 67, 68
prediction of quasiternary section of quaternary phase diagram 99-102
Si-Al-O-N system and prediction of quasiternary section of quaternary phase diagram 99-102
Al_2O_3 105
complex systems 21, 22
and nuclear reactor accidents 137, 139-42, 144
prediction of quasiternary section of quaternary phase diagram 99-102
alloys *see* hot isostatic pressing
hot salt corrosion
pyrometallurgy of Cu-Ni-Fe
solidification paths for multi-component systems
steels
AOD (Argon Oxygen Decarburisation) 132
applications 53-5
alloys *see* hot isostatic pressing
corrosion *see* corrosion resistance loss
high-temperature corrosion
hot salt corrosion
ores *see* pyrometallurgy
steels *see* C potential
computer assisted
continuous casting
corrosion resistance loss
low C stainless steels
see also CVD
EMF
Fe and Cu ions
hot salt corrosion
nuclear reactors
quasiternary section
solidification paths
WC tools
Ar 146
and CVD process 109-10
and high temperature corrosion of SiC 171-2
and low C stainless steels 129-30, 132
Si-W-F(Cl)-H-O-Ar system 109-10
WCl_4-$SiCl_2H_2$-H_2-Ar system 114
austenitic stainless steels: prediction of loss of corrosion resistance in 85-93
see also low C stainless steels

BaO 137, 139-41
BCl_3 115, 116
Bragg-Williams approach 33

C 35
C-Cl-Cr-H-N-Na-O-S system, and hot salt corrosion of superalloys 56, 61-4
complex systems 22, 23
and computer assisted development of high speed steels 70-6
Fe-C-Ca-Mg-N-O-Si matrix for gas-metal-slag system 22-3
Fe-Cr-Mo-W-C system and computer assisted development of high speed steels 70-6
Fe-W-Cr-C system, ZPF lines in 46
and hot salt corrosion of superalloys 67
in interstitial solution 36
and multicomponent diffusion in compound steel 210-14
Ni-W-C system and calculated phase

diagrams 79-83
potential during heat-treatment of steel 176-82
C activity in industrial furnaces 178-81
C activity of multicomponent steels 181-2
and production of metallurgical grade Si 200-8
steel, multicomponent diffusion in 210-14
ZPF lines 46
see also high-temperature corrosion of SiC
low C stainless steels
prediction of loss of corrosion resistance
WC tools with Co-Fe-Ni binder phase
CH_2 168, 174
CH_3OH 179, 181
CH_4 168, 173-4
Ca 146, 156
and complex systems 22-3
and continuous casting process 184
Fe-C-Ca-Mg-N-O-Si matrix for gas-metal-slag system 22-3
and hot salt corrosion of superalloys 67
liquid solution 40
CaO 23, 105
and nuclear reactor accidents 137, 139-44
and pyrometallurgy of ores 152, 156, 158
CaO-SiO_2-CrO_3 slag system 130-1
Cd: Ag-In-Cd system 140
chemical vapour deposition *see* CVD
ChemSage 111, 122, 127, 207
Cl/Cl_2: C-Cl-Cr-H-N-Na-O-S system, and hot salt corrosion of superalloys 56, 61-4
and hot salt corrosion of superalloys 61-4, 67
liquid solution 38
Si-W-Cl-H system 109
Si-W-F(Cl)-H-O-Ar system 109-10
Clausius-Clapeyron equation, generalised 103-5
clogging prevention in continuous casting process 183-6
Co
Co-Cr-Fe system, ZPF lines in 48-50
Co-Fe-Ni binder phase *see under* WC tools
and high temperature corrosion of SiC 168, 174
and hot salt corrosion of superalloys 67
CO 23
and C potential during heat-treatment of steel 178
low C stainless steels 129-33
and production of metallurgical grade Si 200-8
CO_2 23, 60, 64, 139, 204
complex equilibrium approach: to low C stainless steels production 131-2
to silicon production in electric arc furnace 201-2
complex systems 21-4
computer-assisted development of high-speed steels 70-6
concrete and corium of reactor *see* MCCI
continuous casting process, preventing clogging in 183-6
CORCON 143, 145
corium and concrete of reactor *see* MCCI
corrosion: resistance loss predicted in austenitic stainless steels 85-93
see also high-temperature corrosion
hot salt corrosion
counter-current reactor approach to production of metallurgical grade silicon in electric arc furnace 203-8
Cr
and C potential during heat-treatment of steel 181-2
C-Cl-Cr-H-N-Na-O-S system, and hot salt corrosion of superalloys 56, 61-4
in alloy 22

Co-Cr-Fe system, ZPF lines in 48-50
complex systems 22-3
and computer assisted development of high speed steels 70-6
and continuous casting process 183, 184
Fe-Cr-Mo-W-C system and computer assisted development of high speed steels 70-6
Fe-W-Cr-C system, ZPF lines in 46
and hot salt corrosion of superalloys 61-4, 67, 68
low C stainless steels 129-30
magnetic effects in solution phases 40-1
and multicomponent diffusion in compound steel 209-13
ZPF lines 46
see also prediction of loss of corrosion resistance
Cr_2O_3: and continuous casting process 183-6
and hot salt corrosion of superalloys 56, 60-6
low C stainless steels 129, 131-3
Cu: Cu-H_2SO_4-H_2O subsystem 125-6
EMF evaluation 187-9
and solidification paths for multicomponent systems 97
sub-lattice model 35
see also Fe and Cu ions
pyrometallurgy of Cu-Ni-Fe
Cu-Mn-O-S system 187, 189
Cu-O-S system 188
Cu_2O 188-9
Cu_2S 158, 187-9
Curie temperature 29, 40-1
CVD (chemical vapour deposition) of WSi_2, thermodynamic simulation in 108-17
limitations and further development 114-16
procedure 109-13
results 113-14

databases and programs 111, 116, 143-5, 212-13
for analysis of diffusion in multicomponent phases 198-9
see also ChemSage, SGTE, Thermocalc
Debye function 27
Debye-Hückel function 122
decarburisation 132
see also low C
depleted zone in steel 86-92
differential thermal analysis (DTA) 143
diffusion *see* multicomponent diffusion
DTA (differential thermal analysis) 143
dynamic processes, steady-state calculation for 193-5

electric arc furnace, production of metallurgical grade silicon in 200-8
complex equilibrium approach 201-2
counter-current reactor approach 203-8
stoichiometric reaction approach 200-1
electrolysis 187
electrode potential *see* Fe and Cu ions
EMF evaluation from potential phase diagram for quaternary system 187-92
enthalpy 17, 26-7, 51-2, 194
entropy 17, 26-8, 51-2
equilibrium 18, 21
gas-salt 60-1
and hot isostatic pressing of Al-Ni alloys, estimative treatment of 105-6
see also graphical representations of equilibria
phases and phase diagrams
estimative treatment *see* hot isostatic pressing
eutetic reaction in hot isostatic pressing 105-6

F in Si-W-F(Cl)-H-O-Ar system 109-10
Faraday's constant 18, 188
Fe: alloy 22
Co-Cr-Fe system, ZPF lines in 48-50
in Co-Fe-Ni binder phase *see* WC tools with Co-Fe-Ni binder phase

and Cu ions in aqueous H_2SO_4 solutions as function of electrode potential 118-28
Cu-Fe-H_2SO_4-H_20 complete system 122-6
Cu-H_2SO_4-H_20 subsystem 120-2
Fe-H_2SO_4-H_20 subsytem 119-20
further developments 126-7
heat capacity of 29-30
in interstitial solution 36
liquid solution 39
low C stainless steels 129-30
magnetic effects in solution phases 40
and nuclear reactor accidents 137
P-T phase diagram for 32
see also pyrometallurgy of Cu-Ni-Fe steels
Fe-C-Ca-Mg-N-O-Si matrix for gas-metal-slag system 22-3
Fe-Cr-Mo-W-C system 70-6
Fe-Cr-Ni-C alloys *see* low C stainless steels
Fe-H_2SO_4-H_2O subsystem 126-7
FeO 23, 152, 156, 158
Fe_2O_3 23, 39, 158
Fick's law 196
furnaces, industrial: steady-state calculation for dynamic processes 193-5
see also C potential during heat-treatment of steel
electric arc furnace

gas: gas-metal-slag system matrix 22-3
gas-salt equilibrium 60-1
interaction with Cr_2O_3 61-5
hot-gas corrosion *see* high-temperature corrosion phase 38
and CVD process 108-10, 114-15
low C stainless steels 129-30, 132
and pyrometallurgy of Cu-Ni-Fe sulphide ores 151-3
and Si production in electric arc furnace 202
Gibbs energies 14, 191
and basic thermochemical relationships 17-19, 21-4
and hot isostatic pressing 103
and hot salt corrosion 60
low C stainless steels 133
mathematical relationships 51-2
models and data 25-6
and multicomponent diffusion 199
and nuclear reactor accidents 137-9, 144
for pure stoichiometric substances 27-32
and solidification paths for multicomponent systems 94
for solution phases 33-41
ideal gas 36-7
interstitial 35-6
liquid 37-40
magnetic effects 31, 40-1
sublattice model 34-5, 40
substitutional 33-4
Gibbs-Duhem equation 17, 51-2, 104-5
Gibbs-Helmholtz reaction 27
graphical representations of equilibria 43-50
property diagrams 49-50
special cases 46-9
Zero-Phase-Fraction lines 47-8
Gulliver-Scheil's model 94, 97

H/H_2
and C potential during heat-treatment of steel 178
C-Cl-Cr-H-N-Na-O-S system, and hot salt corrosion of superalloys 56, 61-4
and CVD process 109, 115, 116
and hot salt corrosion of superalloys 61-4, 67
ideal gas 37, 38
Si-W-Cl-H system 109
Si-W-F-H system 109
Si-W-F(Cl)-H-O-Ar system 109-10
WCl_4-$SiCl_2H_2$-H_2-Ar system 114
see also high-temperature corrosion of SiC
HCl 60-1, 65
H_2O 20-1
Cu-H_2SO_4-H_2O subsystem 125-6

and CVD process 109-10
and high temperature corrosion of SiC 168, 170-1
and hot salt corrosion of superalloys 60-1, 64, 65, 68
ideal gas 37
and nuclear reactor accidents 139
see also Fe and Cu ions
H_2SO_4, aqueous *see* Fe and Cu ions
Helmholtz energy 17, 26
high-temperature corrosion of SiC in H-O environment 163-75
Si-C-H systems 167-8
Si-C-O-H ssytem 170-1
thermodynamic analysis 167-8
high-speed steels, computer-assisted development of 70-6
hot isostatic pressing of Al-Ni alloys, estimative treatment of 103-7
Clausius-Clapeyron equation, generalised 103, 104-5
and equilibrium 105-6
hot salt corrosion of superalloys 56-69
data used for calculations 56-9
extension to higher order systems 67
future developments 67-8
gas-salt equilibrium 60-1
interaction of gas and salt with Cr_2O_3 61-5
limitations of data and calculated results 65-7
hot-gas corrosion *see* high-temperature corrosion

ideal gas, Gibbs energies for 36-7
In: Ag-In-Cd system 140
internal energy 17, 26, 51-2
interstitial solutions, Gibbs energies for 35-6
ionic solutions 37-40, 118-28
see also Fe and Cu ions

K 38, 67
KCl 37, 38
kinetics 191
steady-state calculation for dynamic processes 193-5
see also multicomponent diffusion

Lagrangian multipliers 22, 51-2
La_2O_3 137, 139-41
Legendre transformation 18, 51-2
liquid: and hot isostatic pressing 103-5
solutions, Gibbs energies for 37-40
low C stainless steels, thermochemical conditions for production of 129-34
complex equilibrium approach 131-2
engineering conclusions 132-3
mass action law approach 129-31

M_6C carbide 72, 77-82
$M_{23}C_6$ carbides 72, 92, 129, 133
macroscopic process 199
steady-state calculation for dynamic processes 193-5
magnetic effects and Gibbs energies for solution phases 31, 40-1
marine environments *see* hot salt corrosion
mass action law approach to low C stainless steels production 129-31
mathematical relationships between Gibbs energies and other information 51-2
matte and pyrometallurgy of Cu-Ni-Fe sulphide ores 151-2, 157-60
blowing 152-4
phase separation 154-5
maximum partial pressure surfaces 168-74
Maxwell relationship 17, 51-2
MCCI (molten core-concrete interaction) model of nuclear reactor accidents 137, 139-40
development of model 139-40
evolution of different phases during progression of 143-4
solidus-liquidus temperatures for compositions appropriate to 140-3
thermodynamic database used 140-6
thermodynamic model 138-9
vapour phase over oxide system 144-6
Mg 144, 146

Al-Mg-Si solidification paths 94-8
complex systems 22-3
Fe-C-Ca-Mg-N-O-Si matrix for gas-metal-slag system 22-3
and hot salt corrosion of superalloys 67, 68
liquid solution 39
solidification paths 94-8
MgO 23, 137, 139-41, 144
Mg_2Si 95, 97-8
microscopic process 191
see also multicomponent diffusion
Mn: and C potential during heat-treatment of steel 181-2
and continuous casting process 184-6
Cu-Mn-O-S and Mn-O-S systems 187-9
in interstitial solution 36
and multicomponent diffusion in compound steel 210, 211, 213
Mo: and C potential during heat-treatment of steel 181-2
and computer assisted development of high speed steels 70-6
and corrosion resistance 91
Fe-Cr-Mo-W-C system and computer assisted development of high speed steels 70-6
and multicomponent diffusion in compound steel 210, 213
models and data 25-42
see also Gibbs energies
molten core-concrete interaction *see* MCCI
MPPS (maximum partial pressure surfaces) 168-74
MTDATA 144
multicomponent diffusion phases 196-9
in compound steel 209-14
C steel and alloy steel 210-14
stainless and tempering steel 209-10
database for analysis 198-9
phenomenological treatment 196-8
see also phenomenological treatment
multicomponent steels, C activity of 181-2
multicomponent systems *see* solidification paths

N/N_2 23
and C potential during heat-treatment of steel 178-81
complex systems 23
and complex systems 22-3
Fe-C-Ca-Mg-N-O-Si matrix for gas-metal-slag system 22-3
and hot salt corrosion of superalloys 60-4, 67
prediction of quasiternary section of quaternary phase diagram 99-102
and pyrometallurgy of ores 156, 158
Si-Al-O-N system and prediction of quasiternary section of quaternary phase diagram 99-102
Na 56, 61-4, 67, 68
NaCl 56, 57, 60-1, 65
Na_2CrO_4 57-9, 64-6
NaOH 57, 58, 59, 60-1, 65
Na_2SO_4 56-9, 60-1, 64-6, 67
Neel temperature 29, 40-1
Nernst equation 119, 125-7
Ni 20, 29
Al-Ni alloys *see under* hot isostatic pressing
in Co-Fe-Ni binder phase *see* WC tools with Co-Fe-Ni binder phase
and hot salt corrosion of superalloys 67, 68
low C stainless steels 129-33
and multicomponent diffusion in compound steel 209-13
Ni-W-C system and calculated phase diagrams 79-83
see also prediction of loss of corrosion resistance
pyrometallurgy of Cu-Ni-Fe
NiO 19, 20
NiS and Ni_3S_2 19, 158
Noranda process 154-5
nuclear reactors, severe accidents in, phase equilibrium calculations and analysis of 135-50
see also MCCI

O/O_2 20, 146
and C potential during heat-treatment of steel 178-81
C-Cl-Cr-H-N-Na-O-S system, and hot salt corrosion of superalloys 56, 61-4
basic thermochemical relationships 19, 21, 23
complex systems 22-3
and continuous casting process 184
Cu-Mn-O-S, Cu-O-S Mn-O-S systems 187-9
and CVD process 109-10
Fe-C-Ca-Mg-N-O-Si matrix for gas-metal-slag system 22-3
heat capacity of 28
and hot salt corrosion of superalloys 60, 61-4, 65, 67, 68
ideal gas 37, 38
liquid solution 39
low C stainless steels 129-30, 131, 132
prediction of quasiternary section of quaternary phase diagram 99-102
and pyrometallurgy of ores 153, 156, 158
Si-Al-O-N system and prediction of quasiternary section of quaternary phase diagram 99-102
Si-W-F(Cl)-H-O-Ar system 109-10
see also high-temperature corrosion of SiC
one-dimensional phase 48-9, 202

P 67
Parrot program 212-13
partial equilibrium between C steel and alloy steel 210-14
peritectic reaction in hot isostatic pressing 105-6
phases and phase diagrams: and basic thermochemical relationships 19-20, 29
calculated *see* nuclear reactors
WC tools with Co-Fe-Ni binder phase
and computer-assisted development of high-speed steels 71
and CVD process 108-15
and EMF 187-92
and graphical representation of equilibria 43-9, 51-2
and hot isostatic pressing 103, 106
and hot salt corrosion 58-9, 66
and low C stainless steels 130, 131
and pyrometallurgy of Cu-Ni-Fe sulphide ores 151-8
for quaternary system, potential, EMF evaluation from 187-92
separation in matte 154-5
and Si production in electric arc furnace 202
and solidification paths for multicomponent system, calculation of 95-6
phenomenological treatment of diffusion in multicomponent phases 196-8
Pitzer model 122
poly-Si 108
potential phase diagram for quaternary system, EMF evaluation from 187-92
prediction
of loss of corrosion resistance in austenitic stainless steels 85-93
of quasiternary section of quaternary phase diagram 99-102
pressure *see* hot isostatic pressing
pressurized water reactors *see* nuclear reactors
process simulation, towards 191
see also electric arc
multicomponent diffusion
steady-state calculation
programs *see* databases and programs
PWRs (Pressurized Water Reactors) *see* nuclear reactors
pyrometallurgy of Cu-Ni-Fe sulphide ores (matte, slag, alloy and gas phases) 151-62
blowing matte 152-4
phase separation in matte 154-5
solidification and recrystallization 155-8

thermodynamic models and data 158-9

quasiternary section of quaternary phase diagram, prediction of 99-102
quaternary system, potential, EMF evaluation from 187-92

recrystallization and pyrometallurgy of Cu-Ni-Fe sulphide ores 155-8
Redlich-Kister series 33, 139

S 20
C-Cl-Cr-H-N-Na-O-S system, and hot salt corrosion of superalloys 56, 61-4
Cu-Mn-O-S, Cu-O-S Mn-O-S systems 187-9
and hot salt corrosion of superalloys 61-4, 67
sulphide ores *see* pyrometallurgy of Cu-Ni-Fe
SO_2 19, 21, 60, 65
SO_3 60-1, 67
salt corrosion *see* hot salt corrosion
SER (standard element reference) 28, 138, 179, 181
SGTE (Scientific Group Thermodata Europe) 14
and C potential 178
and computer assisted development of steels 73
and CVD process 110
and EMF 189
and high-temperature corrosion of SiC 167
and hot salt corrosion 56, 60
member organizations 8
and models and data 29, 31, 41
and nuclear reactor accidents 138
and solidification paths for multicomponent systems 98
Si 146, 156
and C potential during heat-treatment of steel 181-2
Al-Mg-Si solidification paths 94-8
complex systems 22-3
and continuous casting process 184-6
Fe-C-Ca-Mg-N-O-Si matrix for gas-metal-slag system 22-3
and hot salt corrosion of superalloys 67
and multicomponent diffusion in compound steel 210, 211-13
poly-Si 108
production *see* electric arc furnace
solidification paths 94-8
see also high-temperature corrosion of SiC
Si-Al-O-N system and prediction of quasiternary section of quaternary phase diagram 99-102
SiC 202, 204-7
Si-C-H systems 167-8
Si-C-O-H system 170-1
SiH_2Cl_2 109, 114
SiH_4 109, 114, 115
Si_3N_4 99-102
SiO 23, 144
and high temperature corrosion of SiC 168, 174
and production of metallurgical grade Si 201-2, 204-7
SiO_2 23
$CaO-SiO_2-CrO_3$ 130-1
and high temperature corrosion of SiC 168, 170-2
and hot isostatic pressing 105
liquid solution 39-40
and nuclear reactor accidents 137, 139-41, 142, 144
P-T phase diagram for 32
prediction of quasiternary section of quaternary phase diagram 99-102
and production of metallurgical grade Si 200-8
and pyrometallurgy of ores 152, 153, 156, 158
SiO_4^{4-} 40, 101
Si-W-Cl-H system 109
Si-W-Cl-H-O-Ar system 109-10
Si-W-F-H system 109
Si-W-F-H-O-Ar system 109

simulation *see* process simulation
slag: $CaO-SiO_2-CrO_3$ system 130-1
gas-metal-slag system matrix 22-3
low C stainless steels 129-33
and pyrometallurgy of Cu-Ni-Fe sulphide ores 151-3, 157-8
SOLGASMIX 111, 167, 173
solidification:
paths for multicomponent system, calculation of 94-8
phase diagram 95-6
and pyrometallurgy of Cu-Ni-Fe sulphide ores 155-8
solid phases and prediction of quasiternary section of quaternary phase diagram 99-102
solution phases *see under* Gibbs energies
$SrCl_2$ 37, 38
SrO 137, 139-41
staged reactor model 193
stainless steels: multicomponent diffusion in 209-10
see also austenitic stainless steels
low C stainless steels
standard element reference *see* SER
steady-state calculation for dynamic processes 193-5
steels 137
continuous casting process, preventing clogging in 183-6
high-speed, computer-assisted development of 70-6
multicomponent diffusion in 210-14
see also C potential during heat-treatment
stainless steels
stoichiometric reactions and equations 18-21, 23
and CVD process 109-10
Gibbs energies for 27-32
and low C stainless steels 129, 133
in production of metallurgical grade silicon in electric arc furnace 200-1
and Si production in electric arc furnace 202
sublattice model and Gibbs energies for solution phases 34-5
substitutional solutions, Gibbs energies for 33-4
summarising mathematical relationships between Gibbs energies and other information 51-2
superalloys *see* hot salt corrosion

tempering steel, multicomponent diffusion in 209-10
theoretical background *see* graphical representations
mathematical relationships
models and data
thermochemical relationships
Thermocalc: and C potential 177-8, 182
and computer-assisted development of high-speed steels 73, 92
and EMF 189
and solidification paths for multicomponent systems 98, 102
thermochemical relationships, basic 17-24
complex systems 21-4
see also stoichiometric
thermodynamics *see* applications
process simulation
theoretical background
Ti 67
TiB_2 115, 116
$TiCl_4$ 114, 115, 116
$TiSi_2$ 109, 114
two-dimensional phase maps 43-6, 50

univariant phase equilibrium 106
UO_2 135, 137, 139-44
UO_3 144

V
and C potential during heat-treatment of steel 181-2
and hot salt corrosion of superalloys 67, 68
and multicomponent diffusion in compound steel 210, 213
Va 35
and liquid solution 38, 39

Very and Ultra Large Scale Integration (VLSI-ULSI) 108
VOD (Vacuum Oxygen Decarburisation) 132

W: complex systems 22-3
- and computer assisted development of high speed steels 70-6
- Fe-Cr-Mo-W-C system and computer assisted development of high speed steels 70-6
- Fe-W-Cr-C system, ZPF lines in 46
- and multicomponent diffusion in compound steel 210, 213
- Ni-W-C system and calculated phase diagrams 79-83
- Si-W-Cl-H system 109
- Si-W-F-H system 109
- Si-W-F(Cl)-H-O-Ar system 109-10
- W-Si assessed phase diagram 111-12
- *see also* WC tools

WC tools with Co-Fe-Ni binder phase, selection of composition of 77-84
- effects of replacing Co by Fe and Ni 78-81
- favourable C contents of family of alloys 77, 81-3
- region of favourable C contents 78

WCl_6 109
WF_6 109, 112-13
WSi_2 *see* CVD

Y 67

Zero-Phase-Fraction lines (ZPF) 43-8
Zn 35, 105
Zr 67, 146
Zr-ZrO_2 (Zircalloy) 135-44